A fist full of lies

Copyrights in the music industry

B. Jacobs

First edition: November 2010
Second translated edition: December 2010
Third revised edition: June 2011
Fourth translated revised edition: June 2011

ISBN: 978-1-4477-6501-1

Contents

"That pig blood may flow the same as in de bestseller of Stephen King"

Preface

In 2009 I started amongst other things an empirical research on the music industry. This composition is about the music industry and a contingent major breach of copyright. For this, I used objective scientific methods, particularly I made a lot of phone calls to companies. You will have to believe it for yourself, because at the moment I don't think my name should be mentioned on this book, but I am somewhere in my twenties and a trained criminologist. So it's certainly scientific reading. What the research shows also has social relevance, so beyond the scientific target group, I also hope to reach people with less object knowledge and to make them interested in this topic. That's why this is a book and not a publication in a scientific journal. Apart from the fact that finding a publisher looks more like winning a lotto. Before I continue, I should serve you the truth and say that I wouldn't get paid for a publication in a magazine. Just so you know. I may make costs this way for printing the book, but at at least now I have any chance on revenues. Besides that, I doubt wether my results, in any possible scientific form, would be placed in a magazine at all by an editor. It would probably not be good enough: my first work. Besides that, I don't have a client to do this job for. That would also be something that would make an editor of a magazine raise his eyebrows. It's a research only of myself. And don't think that you shouldn't read this book then and shut it immediately if I think that my work can't be published in a professional magazine, because maybe it is the other way around. You should be refusing to read a professional magazine, perhaps that's where the partial truths are in.

Other things you can expect from this book are some personal anecdotes of my own besides the results of my scientific research. Some things in the music industry, unfortunately, strike me personally. More specifically, it means that more or less seperate from my business calls, interviews, questions, emails, etc., I used certain channels that are identified and described in my data, for personal reasons. Those things that are only of my own personal interest, are not mentioned in this publication. That is only of importance to me and not to others. In my opinion it's not necessary to be specific about this. My personal history and background are unrelated to this research. Nevertheless, I permit myself with the freedom I (of course) have as an author, to make personal statements besides my objective findings coming from the data I collected. Several things are the cause of it, which can be found in this book!

"I've called them, I've reached them, we are talking; why must I wait now for an answer?"

1. Calling in the music industry

Let me start by giving my vision on what actually happens in the broad sense when someone calls a business in the music industry or (semi) government agency. Later in this book there is no space to repeatedly write it down or to mention it. You wouldn't be able to read it between the lines. You will however find a chapter amongst others where I wil digress upon the advocacy. I find it important to mention it seperately, because it has social relevance. Perhaps you stop reading after this chapter. It's ok with me, because the most important thing is in this chapter. If you do read on, you may see the rest of this book in the light of the information in this chapter. In any case I would like to tell you the following.

Generally, when you believe that there are certain artists who don't have the copyright and who nevertheless claim it, you will not be put in the right. On the internet at allmusic.com and on the website of BumaStemra information about who is the registered holder of a song, is freely available. Usually it is the singer of the song and/or the band members who play the song. This makes sense, because without a registration at a copyright organization like Buma Stemra, cd's can't be pressed, because that is forbidden for presses. Similar to Buma Stemra is the Recording Industry Association of America (RIAA) in the United States. For these organizations, however, it makes no difference who registers which person as a composer / holder of a song and there is no control on wether someone is actually the maker of a song. The only thing that is randomly checked, is wether the specified song titles match the songs on the cd. This system provides enough reason to doubt wether someone does have the copyright and perhaps the real creators are unknown to the relevant legal authorities (record labels / Buma Stemra). It also gives me a reason to want to know more about how the owner made it. This is an important subject in my research and sometimes something in a conversation to put the focus of the respondent on. Sometimes a hopeless objective to achieve, as people, money earners in the music industry, refuse to deal with that and behave very laconic concerning this. And they don't dare to admit that there could be an aggrieved party. A party that is not involved in music. A party that might consist of a single citizen. Moreover, people in the music industry feel empowered enough to say that a person who declares this, must be crazy. This is about a person about whom songs are. It is something they don't want to face.

By some record companies, a person, who for example says that his privacy and live is being assaulted by a band that makes music and that this band due to reasons including distance, can't be the author, is seen as a dangerous threat. It's just somebody who harasses a legitimate company, whatever the company concerns. One doesn't even need to know his own business. It is about a (psychiatric) individual with a very special mental illness. Often the attendant or the company lawyer, etc. first discusses the case with the other staff and then they draw that conclusion, without that a psychiatrist was involved. That is suddenly accompanied by threats, insults and a first-name way of speaking coming from a person who said earlier to not know anything about it or who would transfer it to another department. Because of all this, I'm sure that if I continued to raise this a bit further at the larger record companies, that I would have been arrested. So with state violence and some government report, because of which I could now still be staying in an institution with some sort of criminal record. All through acts of the involved record company, the police and the whole constitutional state, since it is able to label me as a sort of dangerous terrorist.

But in short, the standpoint is that some artists are just executants and not at all the makers, that they say they are. In the most extreme case, for example, I made the songs myself and processed it into sheet music, of which I've been robbed. In another extreme case, somebody, without my knowing, recorded my spoken words and clues for the music (beats) and instruments with a sound recorder and processed it to a music cd. Also, it is possible that I am part of the reference frame of the creators of music, so I'm offended by the performers of songs. At the same time it concerns private information.

When calling it is usual to get to speak to the reception first. In other cases you get a voicemail with an extensive menu. When calling to the United States this is often the case, even with the larger record companies. The reception connects you through, but often after the receptionist no one can be reached and you can only get for example the voicemail of a person in a superior position. It happenend once that because of that I accidentally informed someone in his voicemail that they should eliminate him. I made an extensive list beforehand of things I wanted to say in English, because I wanted to call the big U.S. based record company 'Koch'. It is about violating copyright law and the international spread of for example something specific, which is taking place through that record company. But I couldn't reach anyone of a record company of those proportions and accompanying status. It should of course not be like that; it is in fact totally unacceptable and plain indecency. There was only a comprehensive answering machine, or maybe only of the person I wanted to speak. I did, after I spoke the receptionist, my fairly pointless story, because no one talked back. What happenend next, is that I got angry and said something like that they should shoot him in his head if he did his job like that. Bending private information into money making music is a bad business as well.

Calling a receptionist is not always pleasant. For example, it is annoying to speak to someone who seems te be uncomfortable. Nine times out of ten, however, one gets either a kind of minor girl ór an about 30 years old (house) wife on the phone. The girl sounds like a pimple squeezed by a penis and the woman sounds like a raped woman. It's no fun to be helped that way. A business conversation deserves a businesslike tone. Someone working for a business should always behave in a professional way. However, the reception is often the entrance to a business. Often, there is something where you need to go through.

Strikingly, the receptionist usually knows and understands nothing, even if clearly explained. From the moment she has contacted a superior, she can even put words in the mouth of the caller that he/she never could have said or intended. If you can reach someone from another department or a manager, then actually the same thing happens. They don't understand it and won't say anything about the way the company conducts business. It also takes a big effort to make the staff say anything about the operation of the business, without that they say something that contradicts it on specific issues a little later. In addition, companies in the music industry shove the responsibility on copyright easily to something, somebody else or caller himself ("He should have registered it"). When they finally understand that they are the ones that are addressed about something, they in fact move from a defense into an attack position. That defense and facade is actually useless anyway if the company doesn't do wrong things that give its workers a bad conscience. Simultaneously, the ignorance of certain employees, including company lawyers is shocking. Think about the process of making music, checkpoints for defining copyrights by Buma Stemra, etc. Probably this causes employees in the music industry to suddenly get mad in a conversation and deliberately continue to miss the point overwhelmingly. To point it out, I've talked to people who answered every question with the same words or people who hung up the phone, because they didn't understand why they should talk to me. Why then answer the phone at all in such a business?

Besides the fact that getting angry, not talking or talking next to the point and having no knowledge of the issues within the company, etc. is unprofessional and brutal, there is also (unsurprisingly) a personal, say human response that is abnormal. Obviously, when something is clearly well explained and it really wasn't in Chinese, it then is shockingly indecent to say that you still don't understand. That is something that applies to every position in the company, even the lowest. In addition, they should show understanding for example for someone who indicates to be in the position of a victim, because there is a band who made songs about his/her life, that are now being published to the world. This human response is lacking. Yet, it is a sad fact, for which compassion is needed in the communication. Doesn't that happen and are they inappropriately compassionate and then offensive, then that is an unacceptable response. It is something that generally is not done in social life.

Despite this sort of objective rejection, it seems that the lady at the reception wants to have sex with you. The inviting bodylanguage will guaranteed show when you would come to the counter. Her way of talking shows that the receptionist seems to think that you think she's horny. In doing so, she'll categorically say "No" to all business related questions. Perhaps it is a sign of the mental deficiency which underlies the malfunctioning of a company?

As a final comment I want to say that the people seem unprofessional. For a caller this can be an offensive experience. Sometimes the receptionist seems to still have a baby somewhere. Or she looks like a tormented pregnant home worker, who can't possibly be who she says she is when she answers the phone.

This was the first part about the *'front door'* of the music industry. I wanted to get rid of a few things here, before I will take on a more serious approach to the effects of my study. Hopefully I succeeded hereby in giving a first general view on the music industry.

2. Approach of the research

The research is an exploratory study which focuses on the music industry as a whole. The main research question is:

1. *"Are the registered copyrights in the music industry right?"*

In that context, it also concerns the questions:

2. *"How do companies and institutions in the music industry function?"* and:

3. *"How are songs made?"*

As part of the research purpose, all in the music industry occurring businesses, individuals and agencies were approached to try to get an insight in the way they function. The study took place in 2009 and 2010 over a span of more than a year. To expose and understand all the facets, the snowball method is used to get from one contact to another. Then the sources, mostly agencies, were exhaustively contacted, until also newly developed questions were answered and the data became saturated, which became evident by the repeated information that was obtained. At the start of the research and for the approach of the resources, the focus was on the last two research questions. This was done with the aim to obtain as much as possible information and to prevent suspicion to arouse in the approached person. After and during this stage of mainly field study and working out the data, also question one is addressed. Through the field research a clear picture of the music industry arises. Furthermore, I will ascertain in this book, inter alia, on the basis of the collected information, why the applicable copyrights in music are correct or not.

Means/methods

A number of different research methods were used for this study. These include interviews, a survey, observations and telephone and email contacts. Face to face, there has been an extensive interview with someone from a record company. A journalist (working in the music industry) was found willing to coöperate in an open phone interview and an appointment was made with a lawyer. In addition, a survey took place to collect the public opinion about the music industry and especially about the first research question. Also, several observations were made. This happenend in places like dance parties, rehearsal rooms and festivals, but also at record companies and magazine publishers. The phone and email was often used to gain information from companies on specific, well prepaired questions and for example to collect information from abroad. Finally, small (internet) films of artists, interviews and performances were watched and a number of music magazines were read and investigated on relevant points for this study. An editor was asked to give a response to some results of the survey for this research. Up to the present no reaction came to this.

Difficulties

The main difficulties encountered are actually already mentioned in the first chapter on calling. Calls to businesses is something that I did a lot for this research. Then often you only get to deal with people with limited knowledge, the telephonist / reception and the like. On the one side the receptionist can refuse to put you through, on the other hand you are often put through to someone who is unavailable. Then one gets the reception again, a beep or a voicemail. Being called back is very unusual, even after leaving a voicemail message this doesn't happen. This is, because these companies do not want to talk to people who, in their perception, are no business partners. At the same time the personnel shows no understanding of substance and they are suspicious. Some people cut the conversation for that reason. A legitimate company obviously doesn't want to harm itself and principally presents itself only in the proper way. So an empoyee will never say automatically that crimes are committed in the company. This makes gathering relevant information difficult. It is also unwise to explicitly articulate the first research question and to ask for proof of certain copyrights. Because this evidence isn't there and it is based on a believe, immediately a kind of miscommunication and a psychological turmoil in the person being called develops, who would then like to end the call. Or you are led to a legal department where you become at loggerheads with, because they turn around who needs to have evidence of what they register. They indicate that if someone at an earlier time officially registered the same song, or send it to himself by registered mail, he can then bring the case to court. Occasionally, such is a case concerning copyright that appears before the courts. The judge then determines based on the postmark or on the registration date of the song at Buma Stemra or at another copyright organization, who is the recipient of the number. Because this is unfair, as it doen't do justice to the making process, there will be elaborated on this further in the research.

3. An open interview with a record company

I started this research with an open interview with a record company. The case when I started this was, that I didn't have more than the hypothesis that, amongst other institutions, record labels don't handle the copyright correctly. I made it an objective to test this hypothesis scientifically, since I'm trained for that after all. The first thing I did, was make a few research questions of that hypothesis and arrange an open interview with a record company. The purpose of this was to learn more about the ins and outs of the music industry. I didn't have any knowledge of that myself at the start of the research. An open interview with someone of a record company offered the best opportunities for a first exploration of my research object. By going there and ask how everything works, I would gain a basic knowledge that would be useful for the rest of the research.

On the internet I *'googled'* a number of record companies. At some point a list with addresses of record companies automatically appeared. These all carried the coolest names and most were focussed on (alternative) rock music. When trying to approach them a lot of companies on the list turned out to be unreachable, probably because they went bankrupt. This is confirmed in the interview. Occasionally, the phone was answered, but the record company didn't appear to exist anymore. Finally I could reach someone of a small record company in the hardcore dance music segment. This woman immediately showed an open attitude for an interview. The appointment for the interview was scheduled two weeks ahead. Twee weeks later I stood there and while enjoying a cup of coffee, I did the interview which took at least one and a half hour. I was permitted to tape it with my sound recorder, but the interview lasted longer than fitted on the tape, so the last part is not recorded. The interview was completely written out and the information processed and classified in the topics below. Some things only relate to record companies in the dance segment of the market. The wordings of the respondent are of course modified, because of classifying the text of the interview into a number of topics.

Topics from the open interview

3.1. Design of the company
The record company of the respondent has an office, in a calm area of a fairly large city, with a central recording studio in it. Apart from that, is the distributor in the neighbourhood of Rotterdam, to sell the music. Also, there's a website on the internet with a lot of downloads available. The company is of the respondent and a companion, who both started it about twenty years ago. There are roughly two distinct functions in the company (or in a record company in general). One director is mainly working on the creative aspect and the other with the financial aspect. The creative aspect overlaps the financial aspect, because everybody decides on the creative concepts. There are about 40 deejay's / producers permanently employed, who

coöperate on making and thinking of new concepts. The financial part includes: the whole book-keeping, the booking of artists, the contracts and its financial handling.

3.2. Ratio to other record companies

Compared to the bigger record companies who are more active in popmusic, like SONY BMG or EMI, this record company is a small one. The bigger record companies are especially commercially active and make more profit because of that. This record company doens't work like that; its employees are much closer to each other and everybody is behind the product of the company. Also, new workers are informed and educated by everybody, until everyone is at the same level. In the harder 'dance' segment, they are leading and quite big, according to the respondent. The company functions in a niche market of the hardcore dance styles. Here, only a handful of record companies operate.

3.2.A. Niche market

Dance music is a broad concept and includes for example the music of dj Tiesto and a lot of what's on the radio. The niche market of the hardcore dance is with six (dutch) record companies a quite small, but specialized and profitable market. There are only record companies in it who already do it for some time. The market is very controlled in the way that people know who is in it and it is very transparent. The companies working in that music style, need to coöperate to keep the market alive. Without a more or less enforced coöperation, a record company in that

21

market has no reason to exist. Everybody working in it, keeps its own identity, but is also working with the other players to promote the same market. It's a form of 'healthy' competition. This is necessary, because the market for the hardcore dance is so small. If a player would outcompete the other players roughly, then the market for it would be gone and only one product would be left available. Then all the other record companies would go bankrupt. They collaborate, for example, on a compilation cd that contains everybody and on parties, so the consumer choice is bigger, more varied and stays interesting.

3.2.B. Major record companies

Bigger record companies operate on a different type of process, making the stereotypical image that people just know nothing and suddenly become a superstar because of them (e.g. through American Idol). That has to do with the fact that they operate in a more accessible genre that sells well. They work indeed with a certain selection of people, because they make a lot of money by that. Contractually speaking, it shows a difference between large and small record companies. The artist is only an executing performer and only earns something of he/she performs. Of the profit of that again, about half of it goes to the record company. Smaller record companies do that less, which is probably why they make less sales. In dance music is not really common to completely dress the artist and do their make up, also the business involved in the interview is too small for that. However, they will look for a sponsor who dresses the artist, as soon as the artist appears in the picture. The clothes, the gig bag, headphones, etc. are then the mark of the sponsor. Smaller record companies earn less of a performance by an artist, because their artists more often (partially) produce their own music. When the artist finds that the record company earns too much of a performance, he/she just goes to another company. It is the eternal debate between the artist and the record company about what will be placed in the contract on the distribution of income from performances.

3.2.C. Causes of bankruptcy of many small record companies

Many small record companies fail to keep their heads above water. The reason why they eventually have to stop it, at first glance looks a bit strange. It doesn't have to do with the creative, but with the financial part of the business. Creativity almost never runs short, but there's too much done with the creative process and too little with the business side. To be more specific, this means that a record company for example produced too much cd's and only sold a few, according to the respondent. It starts with music, but ultimately a piece of administration goes with it. Contracts have to be signed between artists and the company in which the royalty statement is processed that relates to the sales. Also, you have to deal with Buma Stemra. Buma Stemra protects copyright by name registering music and asking fees for publication or reproduction of music of someone else. Often, the case with unprofitable little record companies is, that two or three people, one being the creative brain and/or the artist, simply started it too enthusiastically. Without enough knowledge of what everything costs and what is left at the end of the month. Not enough market research was done beforehand, to see wether and how the product will be sold. The illegal download implies a change in the market. The company needs to adapt to this and continue to do so even if sales were good in the beginning.

Also noteworthy is that while the original label goes bankrupt, artists make their money somewhere else with the same music. They go to another record label. They can also switch if they find that the contract states that the record company makes too much money by them performing. But for small record companies this would be of less importance, because they already earn less on the performance of the artist. Performing is income for an (executing) artist and is of course always possible. For an artist it is actually the main thing, releasing albums has become almost a side issue, because it doesn't sell much anymore. So they go with the same music and act to another record company when it no longer works at their current company.

3.3. Dressing artists / image

Dressing and doing the make up of artists rarely happens at small (dutch) labels. It also isn't usual to do in the dance business. The sponsoring of clothes, etc. of course does happen. The label searches a sponsor. At smaller record companies the imagebuilding is limited to giving advice on how to dress and look sometimes a particular clothing hint. Usually no external business is addressed for this.

At major record labels and more accessible genres it does happen. In the more commercial music the artists are steered and helped in many areas. Think

24

for example about dance teachers, singing lessons and other supporting personnel. The appearance of an artist, clothes, hair, make up, etc., and the image are completely adapted by the record company to fit the project of the producer.

3.4. Concepts

Nowadays in the music industry there can only be thought in concepts. A concept involves more than just a cd. For a new concept, for example, first is thought about which songs get to be on the cd, how many and what it will look like. Next, other things around it are invented, like a party or a merchandiser line. Products have to be included, because only releasing a cd doesn't provide enough income. At the record company, roughly a dozen labels represent the different concepts. The products show what label goes with it, but only connoisseurs hear the differences in the music. In terms of music the labels all seem the same. People who often listen to the music, go to the festivals and keep track of the new music that comes out, hear the difference though. According to the producers the music represents a kind of emotion. There would be no subject in the form of a concrete, specific idea, story or situation from which a concept develops.

3.5. The working method for a song / concept

At the record company of the respondent everybody thinks along about the creative part of everything that is produced. For this, several dozen producers are attached to the company, who further produce for nobody else. As is usual in the dance industry, the producer is often the artist; the deejay. That's also the way it is presented to the world. Sometime is worked with an external singer, who then cowrites the song. What often happens in dance music, is that producers work with samples. This is about pieces of someone else in a record. This can happen in two ways. The first way is to ask permission to use the sample. This is often not a problem. The second way is to use it without permission. The latter probably happens the most.

A record studio is not always linked to a record company. At the company of the respondent there is a central studio in the office of the business. Producers can have there own (private) studio where they make their tracks. They can bring those tracks to the record company, where it will be approved or not. Then is thought about what will be done with it. Next, a master is made in the central recording studio. This master goes to the press. But it can also go completely external, if the producer is so professional to deliver an off-the-shelf product.

3.6. The dance market and its share
In the dance market, the company has a leading position, but it is not hugely profitable. To illustrate this, it's a fact that in a production of 1500 cd's, only 500 cd's generate a profit. This makes thinking of broader concepts necessary, organizing festivals and working with record companies in the same industry.

3.7. The coöperation between record companies
A record company doesn't always work completely independent. Occasionally, an external singer can be asked for a song, who cowrites with the producers and

gets paid for her contribution. In addition, record companies in the hardcore dance music are more or less forced to coöperate with each other, because of the fact that they work in a niche market. Sometimes that is involuntarily when samples are used of someone who hasn't granted any permission. Other forms of coöperation includes parties of several different labels, compilation cd's and other promotional activities for the industry as whole, for which everyone comes around the table. The coöperation between record companies is often evident through their websites. There, they show under the heading '*free for licence*' what other companies could use for a compilation. This is done purely to exchange information among record companies. Despite all this working together, record companies strive to have their own identity.

3.8. Market research
At the business involved in the research it occasionally happens that someone is asked to do a market research; on what people find interesting. A student, for example, will join the company for six months, as part of a final project. It happenend once for the download market in the context of the (illegal) download. At present, however, it is not yet a download market.

3.9. Background of people
Having had any professional musical schooling is not necessary to start a record company or to produce music. But obviously, talent for it, is needed. As for the two directors; they rolled into it without any musical training. The producers group exists of both people who have followed an appropriate training and of

naturals. That means they can sit behind a computer, hold an instrument, can read music, etc. and they automatically manage to make music. New people, who know nothing, get a kind of internal training within the company until they are at the same level.

3.10. Contracts
There's a whole range of contracts possible to make with a record company. In fact, it depends on what both parties would like to have in it. It is necessary, because the copyright has to be (partially) transferred to the record company. In this way, the record company is legally allowed to press music. In addition, an agreement needs to be reached on the financial part of the revenue sharing. What also matters is wether someone is going to stay working for the company or just (occasionally) brings a single song. In the latter case the record company only makes a contract for that product. This doesn't happen relatively much, because it doesn't make enough money. The company has to promote the the product and hence the name of the artist. A song doen't sell itself, one has to do a lot. Withing the company actually everybody is working for one person (the artist). The company gains much more by investing in producers and/or artists who keep working for it for a long period. Compared to small ones, major labels do much more to transform people into artists. This is coupled with contracts that cover many years.

3.11. Idols
'*Idol*'-type programs are a bit of an example of how major record companies actually work, according to the

respondent. Large companies manage it indeed to make someone an artist with a cd and an act if necessary, within a very short period. The artist himself doesn't add anything to the composition of his performance. He's just a sort of called in executant. What is unknown to the audience, is that the participants already signed a big contract before they participate in a qualifying round. So when they win, they are attached to a lot of things. The record company earns a lot, but the artist gets only a tiny percentage of the song sales and can only earn more by means of performances. The income of that will in large part go to those who wrote the song, to the 'Idol' program and to the record company, because the artist is only an executant.

3.12. Promotion through the radio
Many music styles, including hardcore dance, aren't on the Dutch radio, because it is thought of as not suitable for that purpose. It does happen sometimes. That is in a program in which the radio maker thinks outside the box and holds an open attitude to new music styles. For this, there is no need to sign a contract with the radio station. As a promotional activity, record companies can send a new product, a cd to all the dj's and ask if they want to play it. Also, it is the task of the radio dj and the radio producer to look for new things themselves.

3.13. The production of cd's
The decision on the number of cd's to be produced, is always a gamble, but also a matter of product

knowledge. Certain qualities the song has, such as accessibility or experimentality, play a role.

Previously, the sale of cd's resulted in sufficient income for the record company and the proceeds from the performances of the artist weren't necessary. As a result of the illegal download today much less cd's are being sold. From the 90's onwards, it has therefore been necessary for record companies to put something in the contract about this.

3.14. Rippen

The stealing or ripping of music, for example of someone who is not yet with a label, is in the Netherlands difficult to prove, according to the respondent. Apart from name registering the music (e.g. by becoming a member of Buma Stemra), it is possible to prove with a '*master*' in a registered letter that you were first with the song. Ripping indeed happens occasionally, just as using music samples without permission from the owner.

3.15. Why not make your own music without the help of a record company?

It is an impossible task for an artist to financially rely on his music without help from a record company. Most stores will only work with distributors, because they want to do business with a professional channel. A distributor is thus a link between the store and the producer of music. If a self-pressed cd nevertheless ends up in stores, then advertising is needed to distinguish it from the rest. That is basically what the record company is for. They handle the marketing, promotion, financial processing (invoices) and of course the copyright.

3.16. Artists who start their own record company
Successful artists such as the American Dr. Dree, sometimes start their own record company. This way they have more income, because they make so much music and no longer miss the percentage of the proceeds that goes to the record company. Above average creative talent to make music is apparently an essentail feature to start a (small) record company with. In this context, it is worth mentioning that there are record companies that are founded by one or two people who manage to make all the music for their (future) artists.

3.17. The truth in the media
In the media, artists make clear, hard statements about their production, in the way of their background and skills. For example, that they have written their music and lyrics in a certain situation. However, it is not 100% sure wether something is really, and by who himself was made. Just like top athletes, artists get media training about what they should and should not say. (Music) journalists will benefit by playing this game along with the artists, if they don't want to risk losing their jobs. It is socially unacceptable to say that artists don't make their own music if that is true. Because it is unacceptable, people don't want to know the truth that for example a popstar is made. He/she can not simply be replaced by another person. For major labels and their artists, this may mean that they can achieve large profits by long-term investments in a

handful of the same artists. With this method, investing in new talent can mean a big financial risk or costs.

Side note

This interview basicly went successful. However, I must make a comment. What was said in the record company, is actually the way one should run. This is the way a record company should be functioning. Logically, it is presented that way. Wether the company also works like that, is another question. On the one side, it is clear that the respondent had professional knowledge. On the other hand it is possible, although not explicitly said, that the company works with copyrights that are not of the producers. For example, a producer can bring a cd to this record company, of which he says that he made it himself at home, while that is not the case. In addition, someone could be posing as a producer. This is easy to imagine at this company, because the respondent clearly states that all the producers in the company collaborate on a song. Perhaps then the contribution of some producers is negligible. It was also said that they often work with new people with a basic knowledge of zero. These people are more likely to be executants of a work, than makers of a work. If the record company doesn't act properly, then a large part of the verbal and physical communication was lied. The record company doesn't seem to undertake any action if the copyright is actually violated, although they do indicate that this is socially unacceptable and that reclaiming copyrights (of samples in the Netherlands) via a lawsuit usually fails. Apparently, these and other record companies can contribute to the damage inflicted on people by copyrights that don't add up. To put it mildly, record companies pretend things are much better than they really are.

4. The pop journalist

In June 2009 I did a brief phonc interview with a pop journalist. I had called an editor of a music magazine if I could interview one of their journalists. They asked me to mail the question to them. I did this and then got a phone number of one of their pop journalists. The conversation lasted no more than a quarter of an hour. I indicated that among other things, it says in magazines that the music of certain (always mentioned entirely by name) bands, for example is chilling and heartbreaking. It doesn't say any further why that is. While we at school during language lessons exactly learn how to write a (book) review, so with in-depth content. I also said that among others, the 'Hitkrant' (no employer of the respondent) often writes that artists wrote their own song, while the song comes from a producer. It is also concluded in magazines that the song of an artist is about the person with whom they've for example seen the artist walk along the beach in the period before the release. The journalist wholeheartedly admits this and indicates that this is true. According to him, this is because magazines do not feel like it, they have lack of space and because it is too difficult to do it differently. He himself is from a magazine that does speak of the producers if they are the ones who made it. This they don't distil from social journalism or field research, but from the cd cover and the biography on the cover (so just like a consumer). If an artist says that he made it himself, it gets difficult for the media. You won't find out what's going on or who made it. The artist

wants to be seen as a great artist and therefore acts mysteriously. In the studio itself, no journalists come. The respondent gets agitated when he notices that I find that a bit strange. Artist, according to him, are hard to interview and often just say 'Yes' and 'No'. Interviewers have to dig deep for the truth. Interviews usually take place in a hotel. The address of artists is not traceable through the industry. The journalist believes that is normal, otherwise there are fans at the door. (Addresses and such can nevertheless be found using certain websites for finding people.) The name of artists is often, at the beginning of their career (starting bands are in the focus of many magazines), still their own name. This is especially when they are less well-known. He also indicates that everyone want a piece of the pie from the music, such as journalists, magazines, record companies, etc. in the music industry. Hence, many things (are written that) are not correct or not sorted out well. The journalist knows that many artist work together with the same producer who is capable to write their songs. These producers are powerful (a sort of Superman), because they make a lot of songs in relation to other people or companies. Thereby, these producers are not bound to a record company, they operate entirely indepently. The journalist finds it hard to put this in perspective and therefore seems to think of it as just a normal person. As an example he mentions Henk Jan Smit and that he is on almost every cd of the artists of 'Idols'. The program 'Idols' according to him, makes use of a working formula for success. Boni M used female singers who sang behind the scenes. According to the respondent that might be interesting for me as a criminologist to figure out, because that is not right according to him. As far as the journalist knows, recording sounds in public or in private without notifying the person involved, is not punishable and nothing radical could be done with it. I indicated that this is for example a way for people to produce a song (lyrics) about someone. Shortly after that he suddenly twisted the subject to myself by suddenly muttering in the middle of the conversation something unintelligible what sounded like: "..if you would say you heard voices..". I could not confirm this and it is not surprising that the conversation then quickly came to an end. If someone gives a similar twist to the conversation, it becomes very difficult to make him say anything of more substance about the music industry. I suppose it is good for him to distract critics away from the music industry by declaring them a little bit crazy. But it doesn't fill any information gaps that arise due to a malfunctioning of the media. A critical attitude is necessary in the context of criminological research in any legal industry. A more critical stance of pop journalists themselves would be beneficial to their functioning. Also noting down sources can give extra meaning to a text. This is related to finding out and publishing about what really happenend. This means perhaps trying anyway to get in the studio occasionally.

Mail to a music magazine

Halfway through my research, a couple of things made me ask some more questions to editors of magazines that are active in music. I had done my survey and somewhat worked out my data and I wanted a reaction on it from them. Unfortunately, that wasn't successful. It is difficult, because often no contact information, phone number, name, author and no Chamber of Commerce number is mentioned in magazines, texts and websites. One international(!) magazine could only be reached on a mobile number with a non personalized voicemail. The name of the magazine, the organisation or the people involved weren't mentioned. They didn't call me back either. From another magazine (group) I got someone from the telephone reception who asked me to write my questions in a mail and send it to a specific employee. Unfortunately, I got, even on a repeated request, to date no response. So I send an email with a request for comments on my raw survey results. There was also an introduction with a neat, brief description of the subject of the survey, a description of the respondents, the location and the method used (phonebook and beach). In the mail, the following statistics were mentioned:

- Nearly 70% of the respondents consider it unlikely that a 15 year old writes and publishes a hit (song), without outside help.
- Over 60% minds it if artists would say they write their own songs, while that is not true. Also, a majority of 60% thinks it doesn't matter for the cd

and concert ticket sales, if artists lie or not about that fact.

- Nearly 70% thinks it is possible that an artist is unknown with the maker(s) of the song. Of the 30% who think that is impossible, a quarter then doesn't know how an artist is stopped from bringing out a song if he possesses a song anyway, of which he doesn't know who made it. The other people think that an artist is stopped due to a lack of interest, by (a fine of) BREIN / Buma Stemra, by the record company or in a lawsuit by the holders of the copyright or that he then isn't stopped after all.
- Finally, more than half of all respondents suspect that there could be censorship in publications on popular music.

My question:

"Why don't you as pop journalists explore this any further? By this I mean that you, as a concerned media (e.g. based on these results) should do more and better research into the music production process and the agencies involved: record company, Buma Stemra, artist interviews and possibly even further. It seems after all that artists / record companies only have to do with the final process of making music. And how does Buma Stemra work? Buma Stemra doesn't maintain / control in popular music the copyrights at all. From a telephone conversation with Buma Stemra it shows that they '*believe*' this when an artist says he has the copyright and they also don't fine, while people seem to think that.

It appears that editors of magazines are very hard to reach. Employees seem to work isolated in the jungle behind a dead end road. Perhaps this non-response is a tacit proof that they as a matter of fact indeed do socially unacceptable work. Later on, I tried again to call another editor of a music magazine, without explicitly mentioning my research findings. After several failing phone numbers I got an editor of a Dutch rock magazine. This was a middle aged man or younger. Occasion for the conversation: Magazines only write about the (cd) publication, how the concert was and when and where that is. Do they do that too about the working process between the release of cd's? Do they e.g. call on artists in the middle of separate work processes? Are there any aspects that perhaps can be classified some more? No, this is something that isn't necessary according to the young man. The key question of the interview: "Do you also publish it when someone comes to you and says he is offended by the lyrics, music or methods used by a band that has published something?" And I illustrate the case that someone wants to get rid of something in their magazine about the music of the band of his/her neighbor and his methods, because he makes music about his neighbor. No, he doesn't do that interview.

The reason he brings is, because he is just a fierce hardrockfan and to him it's only about the music. Backgrounds or elaborating more on the music is not important, then. Not even that the singer of the band who everyone adores, is lying about the source where the fierce content of his song comes out. So I informed him that it might be of social interest, as there probably are many people who would be interested and that he should do his best as a media to disseminate information that is correct. But this man confirms once more that he feels there is absolutely no need to write about this in a music magazine. The man decides this approximately single-handedly for the whole social constitutional state, as an editor.

5. Buma Stemra and other copyright organizations

On the website of Buma Stemra (comparable to the FIAA in the United States) much information about this organisation can be found. The idea is that music authors transfer the exploitation of the music copyright to Buma Stemra. Buma Stemra then checks where and when their music is used and collects money there. Based on that, the authors get a compensation, often through their label or legal representation. It is impossible for music authors to determine themselves at what time, where and by whom their work is played and go and collect money there. Provided that someone makes for 100% his own music, the withheld administration costs, income for Buma Stemra is very low. According to Buma Stemra an individual can never organize his own interests for these costs; the collective concept is their great strength. The organization consist roughly of the part Buma and the part Stemra. Those who want to publish music, pay a fee via Buma. This includes music in shops, pubs and live performances. The height of the fee is determined by a calculation with the extent of the use. The publisher had to specify to Buma: the number of people that hear the music, how often the music is played and the size of the room where the music can be heard. Who wants to reproduce music (on image or sound carriers) for public use, needs prior permission from Stemra. Think of record companies, importers, broadcasters, producers of background music and advertisers. Then the commercial interest is important in determining the height of the fee. The price one has to pay to Stemra depends inter alia on wether numbers are covered (this is more expensive) or not, the circulation of a sound carrier, the price and the distribution area. The online calculater shows that in most cases someone pays less than a euro to Stemra per pressed cd.

Research by phone

For this research, I called several times with Buma Stemra. They are easy to reach by phone and the staff are quite willing to provide information about what their organization does. Of one employee, I even recieved forms by mail regarding requesting permission from Stemra to produce cd's and a list of addresses of Dutch based manufacturers of sound carriers and intermediaries. Apparently this person thought I was an artist who wanted to release records. To reproduce music by making cd's, one has to arrange the mechanical reproduction rights with Stemra or through an intermediary who has an agreement with Buma Stemra. The staff is not always clear. They go to show what their process is, but they can't explain why they do it like that or ground that their method is correct. For example, they indicate there is a cost advantage if you own the copyrights of a first cd, are not yet registered and sign in later. Substancially seen, there is essentially no difference that could be a reason for a difference in costs, except a different time of paying the registration fee. According to Buma Stemra copyrights can also be registered and protected at a tax agency instead of with them. Buma Stemra can figure out (through the title of the used song) who made something, so that authorization can be obtained for a cd with a reproduction. By mail, I requested of a few songs who its composers are. I recieved an email back in which for each song the holder(s) of the copyright were mentioned and their phone number. In more than 90% of the cases these were not persons, but holders organizations like 'Universal', 'Warner Brothers Music' and 'Emi Music Publishing Holland' or something similar, many at the same address (shown by further investigation). Registering and obtaining permission to make cd's one gets by sending in the title of the songs with the composer(s) and lyricist(s). Random sample checks are carried out by Buma Stemra by asking for cd's and verifying wether the titles on the cd match the registered titles. The objective of Buma Stemra is to collect money and they pay this to the legal holders of copyright, such as record companies. The record companies pay their artists with this. This is regulated in contracts with composers and/or lyricists. Composers / lyricists can also be registered directly and get paid straight from Buma Stemra. In a copyright dispute, the owners are informed by Buma Stemra. This is only possible if both of them are registered in their system with a song. They must then themselves take legal action against each other. Cd presses work through Buma Stemra with licenses. This works as follows. If someone wants to exploit through Buma Stemra and wants to reproduce a cd with a number of songs, he must request their permission by use of a specific request form. Then there is always something of Buma Stemra on the cd. On the form, the following information has to be given:

- Applicant details
- Manufacturer (press) details
- Information of the production:
    * Amount
    * Label
    * Consumer price
    * Area of distribution
- Type of sound carrier:
    * CD
    * Vinyl
    * Cassette
    * Other
- Backordered or first request
  (release date?)
- The material of the sound carrier is from:
    * Studio (studio details)
    * Taken from radio/TV/CD
- Client details (if any)
- Place, date and signature of the applicant

On a separate form:

- The listings of the works on the sound carrier;
    * Title
    * Duration
    * Composer
    * Text poët
    * Music publisher
    * Executant / Artist

It's not necessary to send along a copy of the work and neither anything that serves as proof that the makers actually did make it. There is only the random sample check to see wether their entry matches the registered titles on the product, so mentioned on the cd. It could therefore very well be that the work of Buma Stemra is irrelevant and that they blunder massively. When questioned about this, an employee indicates that it is a matter of belief and faith that when registering the correct title and composer / lyricist are given. According to them it has nothing to do with naivety. There is no control, but they think that if it isn't correct, the other artist of whom the (whole) song is, would take legal action. With for example as evidence a registered letter with the song. Otherwise this person would be very stupid; by not registering his work. They don't understand what else could be going on. Only with a proof of registration you can get a lawyer. Remarkably, the staff very often automatically think that I'm an artist, although I never said that.

Internet / Tax

On the internet it was posed that there was a investigation department for Buma Stemra at tax agencies. After having called a tax office, it appears that this service is unknown to them and they also register nothing, besides documents for loans. They refer to the copyright manifestation foundation and auteursrechten.nl (copyright) on the internet to register. This site is informatively very bad. The information of the tax agency was rather unreliable, because what was said was contradictory or things that were said were corrected at a later point in time. At allmusic.com are songs with its registered holders. Also, you can find registered holders for tunes with different instruments on the internet.

Allmusic.com

Buma Stemra emphasizes the distinction between the executants and the composers behind the scenes, who also need to be protected. Data from allmusic.com demonstrates that this is nevertheless often the same person. Not the real, but the artist name is listed. Sometimes the composer isn't listed or is untraceable. An artist with a lot of songs, like Madonna or Tupac often work(s/ed) with multiple, different and/or well known composers. Famous pop artists often use the same composers. Consequently, the same composer shows up for different artists. So what holds Buma Stemra back to not trace the methods of these composers in making their songs? They are not that many, while their repertoire is enormous. By mentioning the artist name there is the facial recognition of the artist, but not the real name; which is secret. Although Buma Stemra claims only to use real names, the real makers often can't be reached. They are not revealed. In addition, something else is noticed via allmusic.com. When a famous artist is name searched to find his works, then actually all of them are of optimum quality and quite well known. There is no apparent upward trend in the quality of work, because there is not one failure among them. As accounts for Shania Twain for example, there is in addition only a very small repertoire of just a limited number of songs on a few cd's, that she produced in her music career, while it all became quite popular music (in that genre).

Also noteworthy is that e.g. Madonna and Tupac made an exceptional huge amount of songs. They have made these themselves with the help of other composers. The question that then arises is, how and when they have done this, because as an artist publishes more songs, there is less time left to make new music. Think about the gigs, prepairing them, of interviews, promotion and things like that, that need to happen after a song is made. In addition, according to allmusic, Madonna and Tupac worked with many different people per song. For this, they will need to, in between performances, constantly search and lay new contacts with good or promising and perhaps unknown composers, who may live completely somewhere else in the world. Even for someone who has all day to spend or a top secretary, this would be a incredible achievement.

Intellectual Property Hotline

Also the FIOD (Financial Information and Investigation Service) is not familiar with an investigation department for Buma Stemra, they mention the 'Intellectual Property Hotline' in Utrecht. This body seems to incorporate Buma Stemra. They are focussed mainly on trademark counterfeiting of e.g. cd's on the market, clothing and car parts.

BREIN Foundation

The website of this foundation was off the air because of the '*Pirate Bay*' case. Probably, because the exact cause was unknown. It was said that there worked someone who had worked for Buma Stemra, but I didn't speak to this person. They fight piracy of cd's, software and games, large-scale copying and illegal download. I indicated that their work certainly is a great gesture to the holders of copyright (e.g. Pink, Michael Jackson), since they are already very rich and this way they get unsolicited their/more money. Why are they doing this? They keep the already large profit and the financial flows to the artists intact. Buma Stemra is one of their clients. Its financiers are: Buma Stemra, NVPI (music movies), .. . Then the (fanatical) woman on the phone gets angry when I ask these questions and ends the call.

BV POP, part of FNV KIEM

From Buma Stemra I got the details of BV POP. They are available one afternoon a week. It is basically just for members, musicians. Their activities are related to the number of members and what members bring in. The employee states that when someone who is not professional in music, suffers from the fact that artists are wrongly entitled as the rightful holders of copyright and nevertheless implement a work, that this person (the victim) should bring in third party assistance. It is, because of a few things according to him, also a police matter. E.g. for lyrics, the recording of sounds in a home bij unauthorized people; it is not just civil. Also, the victim should use a lawyer against the record company and who they say that wrote it.

Résumé; Buma Stemra / BREIN / Intellectual Property Hotline

These agencies don't understand the concept illegal music. It doesn't exist, because it is illogical and stupid. The entitled artist would start a lawsuit. There he can prove that his work was earlier by registered mail. People don't go into a music creation process at all. It is possible to show this by bringing on semi-manufactured articles, notes with texts, emails, computer files, etc. There is no control on wether the registered makers actually did make the songs. Not at all; there's a blind faith. By the various bodies mentioned in this chapter you get referred back to Buma Stemra for the accuracy of the registration of the makers. Buma Stemra then refers to the cd press, because they would only work with licences. But it concerns licences that Buma Stemra delivers. It is an incomprehensible circle reasoning. Nothing is substantiated.

What if there is antisocial behavior in the music industry? If there are criminals with a voice recorder, secrecy in the social environment, victims don't get to know about this, human trafficking, blackmailing with nude pictures; it's always too late, as Buma Stemra allows it to pass on through its licences. Then people are cheated by the record company and the radio and they will suffer from immaterial damage. What if

49

criminals prompt the autocue of artists on TV, using illegal tapes, if there is then '*flooding*' in the media by one subject, unauthorized people in a home, the neighbors are antisocially involved and do not report it and the police won't report it? Aping of someone who is on the street and next claiming copyrights at Buma Stemra? Amateurs who play for fun for example on drums and someone secretly tapes it, (creates a file about something or someone,) makes music and publishes it. It is possible.

The reaction of the staff of the concerned institutions is, that this supposedly is confusing to them; I take out new things. It is unimaginable for them, new and never done before. They have never heard of it and have no understanding of it. According to the pop journalist it could have to do with power, money and a piece of the pie. BREIN then hangs up the phone. The artists, according to the employee, are supposedly justified by Buma Stemra; they arrange that it is right. On the basis of that, they act against illegal production and distribution. BREIN indicates that Buma Stemra maybe can do something to check the copyrights, by developing a control instrument. In general holders of copyright (take a good read in the copyright law in the Code) are advised by these organization to file a lawsuit, for example against the record company, but an (earlier) official registration of a song is necessary for that. The record company doesn't have to prove that it has copyrights. '*I*' have to prove that a song is mine, while I might have provided five different bands from three countries with only the lyrics of their songs. The idea is ridiculous, it is common knowledge how it is and hanging up is also part of it. There is a yawning gap between the claiming of copyrights by artists on the one hand and the subtraction of data about someone (a source of inspiration) as antisocial behavior on the other hand in the criminal world. The Intellectual Property Hotline, Buma Stemra and BREIN are not familiar in that area and don't have any sections for it. The other artist (and a non-musical victim) would have to start a lawsuit against the publisher. So directly a lawsuit instead of sitting on the negotiating table with the concerned record company, which is also a possibility. A lawsuit may also aggravate the injury to a third. Negotiating will therefore be for a victim probably a better option, because it is less intense and can make money. Putting pressure on the case is followed by a denial of the possibility that the problem could exist. I can then go nowhere with the problem and also no more referrals are made. Only the suggestion (of BREIN) is made that Buma Stemra could go check their records.

6. Record companies

Besides phone contact, I corresponded by email with several record companies and music publishers. The companies I've approached include: Bug Music, Universal, Peer Music Holland BV, Warner Brothers Music Holland, Sony Music Netherlands, AT Productions, Koch and EMI Music Publishing Holland BV. The name of these companies is usually mentioned clearly on the cd and also on the internet, for example in the videoclip on Youtube. Besides that, Buma Stemra appointed, at my request, these companies as representing the holders of copyright (publisher) of a number of songs. With the contact I attempted to raise a few things. My general findings on music publishers and their representatives are: They do not discuss the case, show disinterest, do not know me and that is a reason not to speak to me; I have no connection with the representation. Especially the verbal interviews give the impression that they are dirty people, with a sick personal and social life. Possibly this is carried on in their work. Due to the nature of their work, they are actually like BumaStemra in fact a sick organization. Employees simply hang the phone for no reason and what's usual is actually referring to the '*legal*' department and giving an email address where I can send it to. Often then there's no reaction to that, so you in turn must contact them again, which they then qualify as '*harassment*'. It is actually precisely the opposite way. The moment that I dialup, the phone receptionist starts bothering me. For example if I would be called back, what didn't happen or if I call back later, because the receptionist has agreed on this with me. Then they see my phone number again and know that I've called multiple times, which they indicate to find annoying. It is also clear that mails for the '*legal*' department are usually not dealt with, or are not dealt with purely by the '*legal*' department, because the receptionist just like that knows to mention details from these letters. After a mail with a piece of text from someone it is a company with an X number of employees who all, including those who scrub the toilets, know the name and address of the person and the content of a story with which he would so-called repeatedly show up with, without that they come up with a substantial reaction. That is extremely indecent. It doesn't exist in a legitimate company that is functioning normally. It has also happenend that they mention the head office in America, because they would have nothing to do with it, since they do nothing at their location. Also they think they need to make an impression by saying that the conversation is taped, probably with the idea that the caller will then hang up. According to some companies that represent the copyright holders, the information of Buma Stemra isn't right, because they themselves have nothing to do with it. I should best figure it out through Google and then contact the management of the concerned artist/band.

One representative of Bug Music was very short with me. I spoke with someone who said I should call back for a certain gentleman who would definitely explain it all to me. This person also said paranoid that "they all sit together". Presumably he was referring here to the various music publishers / representatives. To some

53

extent this is true, looking at the phone number (area code) that Buma Stemra gives, because they all look alike. In addition, they are companies with on one address a wide variety of trade names in circulation. Universal / MCA Music Holland BV / etc. are at the same address plus a holding company; MGB holding BV. Bug Music is in Amsterdam, Warner Brothers Music Holland in Hilversum, Sony Music Netherlands in the media park in Hilversum, Universal in Baarn (near Hilversum), Peer Music Holland BV in Naarden (near Hilversum) while the headquarters is in Belgium, AT Productions is in The Hague and EMI Music Publishing Holland BV is also in Hilversum. Two days later I called them again and spoke to the designated person, who wasn't talkative at all. He said that he and the band mambers had sheet music (yes, quite logical otherwise they can't perform music) and didn't know why he should talk to me. I indicated due to a copyright dispute. He hung up the phone. It was a very short conversation, but he could confirm that he did the representation of the band members like this.

At Sony Music I did have direct contact with someone of the department '*Legal & Publishing*'. This was among other things about lyrics that would fully come from somebody else. This company lawyer asks questions and initially takes the case quite seriously. She indicates that lyrics writers get paid, but there is no music until the the lyrics writer and the melody come together, so when lyrics and music come together. This is necessary for the payment of songwriters and melody maker. How do they know that artists did this themselves and are not just performers? Songs are often

54

made from A to Z in the studio. I obviously didn't do that and I'm also not given the opportunity for it and that ends the case for the record company. Repeatedly I indicate that I don't have proof, because my spoken text was taken unobtrusively and has been send internationally (possibly through the internet) and that several songs were made of it and multiple record companies are involved. The company lawyer noted my phone number and email address. Then in an email she asked for evidence of my registration. I passed on to her: "I don't have this." and explained the situation again; that sounds were illegally tapped from my house. Still, I would need evidence of a registration and also I should go to the publisher for this and not to the record company. The name of it wasn't mentioned, as that would concern information that is freely available. Data from Buma Stemra (a list of copyright holders of several songs was requested) shows indeed that they are not. They are however mentioned as the record company with the song on the internet (website of the band and on Youtube), so basically there is a legally dubious relationship. This I also noted in a subsequent email. If the artists don't have copyrights, the record company is not allowed to handle it at any stage or process whatsoever. The company doesn't check the copyrights, doesn't deal with third parties that indicate to also be a copyright holder and principally don't provide any information about their business processes. In addition, everyone in the company has to be aware of the fact that they massively violate the copyright law if that is true, because they also make music in their studios. Who sees an artist at work, can usually only conclude that it is an executant. For someone who is not a professional in music, it is in addition a hopeless task to know all songs / cd's / artists / record companies or to find out what copyrights have been stolen or what isn't correct. It is just a duty of the record company to commit to the law or if they don't have control, to inform people through the media. It is thereby also about a piece of the pie, about the financial interest of third parties that are unknown to the record company. After all, these thirds can address the record company for their money, because nevertheless that they don't occur in their administration, they can have copyrights on the music that the record company works with and publishes. Data from the Chamber of Commerce shows that in this company there is both a holding / financing company and one more focused on the copyright and music business. So initially it was debatable, or they were at least curious in the legal department, but then the case and their own dealings turned out to be not negotiable. By email I further showed them that I need evidence from them of what they state about claiming copyrights and/or information concerning their specific dealings to demonstrate they are negligent and therefore need to contact third party (authors) about copyright. After all, they are the ones that claim copyright on something. By phone the lawyer then unfortunately points out that it is normal that a company doesn't give insight on the ins and outs of its business and refers to data from the Chamber of Commerce for that. So? What is the relationship between the artist and the record company then? Business secret. No copyrights and no relationship with artists? Come on. All artists are actually on their website. If songs are made from A to Z in their studio, then they can never believe by providing these services, to acquire rights on money and shift the responsibility for these activities to the so called copyright holders, a subsidiary company, a cd press or a publisher.

For another record company I had to put my story on email, because the person responsible for the concerned songs was in a meeting. I stated the case as follows. On the radio I heard a couple of songs of which the lyrics are illegally taped conversations and arrangements of that of myself at home in the private sphere. Although I had reason to seriously suspect that besides me there also was recording equipment present, the technical system with which I was bugged was untraceable. Given these facts, the record company is in my opinion not entitled to sign record contracts and also it seems that they nevertheless do it on a large scale. The burden to provide proof concerning copyright lies with them, because they claim them in fact and I don't. I'm unknowingly a part of their business and find that they should (have) involve(d) me more for these songs. So I propose to solve the matter directly with them, without the intervention of third parties, e.g. with an acknowledgement from them in the media that they made a mistake and an appropriate financial compensation. If they fail to try to contact me, it is a serious omission and therefore I will then take legal or other measures against them.

After having called them again wether they recieved it, I recieved a confirmation and it would have been forwarded to the publishing department. However, I never got a response, so I called again after a few weeks. Then I got an email with in the subject line a woman with my name. This is obviously not the subject, but the recipient. They want to inform me then that they do not recognize themselves in my claims to those numbers. This is a paranormal statement, because what is it it based on? They only have half a book-keeping. Also they believe that the claim was unclear and lacks any foundation. That's why they don't feel addressed to deal with my claim any further and ask me to not get in touch with them again. This is signed with kind regards, not by a department, not by a person, but with the name of the record company, Universal BV. They apparently take weeks to simply dispatch the case.

By mail I reply that I firstly don't know about what claim they are talking. I never claimed anything, I stated that they claimed something that's mine. If they don't recognize themselves in that claim of someone about copyrights, they should get in touch with that person that claims it from outside the record company. If they would indeed solve the case, they would never have the problem. In short, the device is: legally handle it. But there belongs a side note. If a lawyer says something about it, it doesn't mean that it has been legally dealt with. A jurist can also as a matter of fact dispose the case, without treating it substantially. By the way, it is not necessarily preferred if the record company qualifies the case as an uptight legal conflict. The negotiating options are then limited. It would be better if an employee of the company, who is slightly higher on the ladder, such as a manager, a representative or someone from sales, picks up the case seriously, so substantially settles it and tries to mediate the case with those involved. So also with people he might not know.

They should be open to this. Secondly, to what contact do they refer when they request that I don't get in touch with them anymore? After all, they have never answered one contact normally. And seeing it is a legal case, I have to seek for contact with them. I am strongly amazed that they don't recognize themselves in my writing. The fact is that they have no control whatsoever over who may have the copyright of the products that are made with their equipment. And they ignore the fact that they constantly braodcast the most simply things internationally, without one responsible interview in between. I find this odd behavior, as the artist with the registered composer come to them with a number of ready-made songs and they merely put it on cd and and multiply it. In addition, they never place vacancies for producers, writers or other composers for their artists, their artists continually say that they write songs alone, are since hit one on the stage and from then on they poop out a new song every month, always at another record company. So why does the record company insinuate that artists make their music themselves? On what do they base that? How and when do artists do that? Making is a different craft from performing. There is a clear distinction to make. So why do they end the case with a threatening email? They deal in copyrights, so they should actually understand my mail. It should be handled by a normal legal department. In addition, the '*creativity*' department should sit around the table with me and give me an official job, because I don't agree with that crap they make. But they probably don't have that department. For messages from other people who casually make music for them, without financial entitlements, they will probably be more susceptible.

I asked EMI Music why the record company refers me to the publisher. They still have something to do with it, don't they? The record company is there for the artists and makes the cd's. The publisher is there for the writers and the composers, so for the copyright holders. Often these are the same people, so yes, then they are enrolled in both institutions. I had to send an email and then she would search the songs it concerned. So I send an email again about the copyrights of a band of which I heard from Buma Stemra that they as a publisher look after the interests of the rightholders. Because it is no longer my first contact with an authority in the music business, I indicated that there is already immediately a small contradiction in that. Previous conversations with record companies and publishers show that they hold a defensive attitude for no reason and they actually don't talk to people who are not registered as holders with them. So in practice, it turns out that there is no question of a concrete representation of something. De status quo doesn't change (nothing). That means that whoever first registers something, always wins it from someone who later says something is his/hers, or somebody who says: "You didn't write that." While it is not earlier as after, for example a without permission recorded conversation, a song is made (with that conversation as the lyrics). The legal framework then apparently works in their advantage, because only the end product can be registered and not lyrics, texts or parts of texts, fragments of music or individual beats, beats on specific instruments, etc. This is at least what employees in the music business believe. Also, the awareness is not present in them to rectify something if they link the wrong composers to a song.

So what needs to happen now? If they don't have a procedure for this, it seems to be the best option for everybody to settle the affairs with the artists who now claim copyright, because it also involves crime committed by third parties. This could include a further indication of who did what and is entitled to what? Can this continue? Who was wrong? Wat kind of a reproduction was made? By who? How? Etc. Obviously it can't go on like this, without this consultation, because someone is always penalized and harassed by people who apparently want to do something with that person's private life, music and media. Such a person is also at risk of becoming a victim of violence, in which stalkers from all over the world play a role. It is coming from the music business so they are responsible for that. Today, as a publisher and an advocate they are not doing it right. They need to base it on something when they say they represent the interests of the right holders. This is what I need to know from them and for this having sheet music is not enough. How do they get it? How was it made? What is the song then about? This is among other things information on how someone can judge wether the executant / copyright holder is indeed the maker. Nowadays, one can easily tell from the artists in the media that they do not know what they are talking about íf they are even talking about their own music. So what ís true? In an appendix I gave my explanation on what the text of the concerned songs is about; the holders will have to do that too if they want to show that they are the makers.

In addition, the band made previous albums of which the content was about me. This was done by people in my surroundings and for this crimes have been committed against me and not been solved, because it is apparently so good to make music. Since it is about my surroundings and the band lives in the United States, and the singer in her lyrics often talks in the third person about herself and in the first person about someone else, she can not have made it herself. That is at least the only way in which the content of the music can be logically understood. Especially if you look at the songtext, texts can describe an entire story. But also the sound of beats and instruments can describe a situation. Each adjective can be imitated by a beat. An instrument can generate a kind of proverbial ideas in people or ideas on how to use it. Likewise, someone who is rocking with an electric guitar shouldn't do a song that secretly is about somebody else. By that a private life is being publicly exposed. It doesn't help then to say that the rocker has made it himself and that it is about something else, because that doesn't change how it really is. An artist simply shouldn't take music that other people have made. If it happens, people don't expect this from a cool guitarist. A couple of things are a serious insult. Also, the band gets better because of it, at least they try to enrich themselves with it and I get worse. Systematically it is formally established like that. It is a copyright conflict involving fundamental rights violation. The publisher / record company will have to do something to withdraw these copyrights and something about their policy to stop future problems. I point out that I can explain this in greater detail and that I would like to see a responsible reaction from them on the short term.

Unfortunately, because once again I get no response, I call them again. As the singer of the band says that she was abused, I ask the representative I get on the phone wether he has proof of that in the form of a doctor's note, of the people involved or of a care facility. If it is not true, it is about fierce lies in the media. In that case, it gets more unlikely that she makes her own music about this past. If the music is really about someone else, then that is appalling. Also, her personal expression in the media then is extremely hypocritical. Apparently, things happen, that are exhibited by the media, that emotionally can't be rectified. The man achieves it to lay the blame on Buma Stemra again and their own department A & R (Artist & Repertoire), who work with the artist. So the man actually denies nothing. According to him they would not be able to control

63

everything, because they have 500.000 songs in their database. If the neighbors would make songs or texts about me or my environment, then there's nothing that could be done about it! But they should check it; it is their business. The man waves it goodbye and starts saying he doesn't understand it again. Who else should check it? Buma Stemra? The media? Journalists? No, they are not required to do so according to him. They don't have to report news! But they may make a report about the concert of yesterday, today and tomorrow and how it was. I point out that they as a publisher make of it that people who come to them with music, made this music themselves. This isn't the most logical option. After all, it can bring a lot of money and status. Anyway, there exist many examples where that suddenly happenend. What also is a sensible theoretical explanation, is of course that many hands make light work and the chance bigger that a big hit is made and that the artist is therefore just an executant. But he doesn't understand all of that. He would forward the mail to someone and the next day this person would contact me. I haven't noticed anything of that.

What also happens when you make a return call later, is that you get to speak a very suspicious operator. At once they can answer everything with a counter-question and even suddenly ask in the middle of the conversation: "*How are you?*". Or they think that I tried something musical and that I'm not good enough. They proclaim that I have too little talent. Communication is then very bumpy. It should of course not be like this and it is not normal. In addition, they regularly yell that they don't creep around in my house and such to steal my music.

64

This, while I never put it like that. I said that someone else has done that and that they do nothing but nevertheless structurally register the copyrights of songs wrongly and work with the music that results from that. The music sort of gets there by itself. Concerning this, the question to an employee why she's actually working there, while it is such a fake corporation with a fake image, is not surprisingly answered with because they earn money there and a number of threats. This is still a fairly ambitious goal, for someone who works in a company that in fact gives shelter to terrorism.

The firm octopus
It is possible to compare record companies with an octopus, because of their long tentacles that reach deep into society. With their tentacles they firstly take away music from people who have nothing to do with the record company and the music industry and they bring it to their artists. With their artists they actually do nothing, except cherish them excessively. This they do in their establishments, similar to the main body of the octopus. Secondly, the same tentacles reach so far into society that they can pick up an individual and throw him/her into a (psychiatric) facility. They do not deal with copyright issues with third parties. They do not handle it and declare someone a fool. If it comes to a lawsuit, then the object of the lawsuit won't be the copyrights, but the mental condition of the one who demands his/her rights. This is possible, because the record company will only say that it isn't right what the other party says and will say about that person that he/she has a mental illness. They will make this the object of the case. They don't have to explain the judge

65

that they don't comply with copyright law. The other party will always lose the case and disappear as a sort of political prisoner in an institution. At the basis of this lies that the record company has a kind of statistical evidence of that the other party, an individual, is dissenting, and therefore not normal, so they start the psychiatry. Their behavior, coupled with the systematic, unexpected, inappropriate remarks like "We don't recognize ourselves in your writing" or "How are you doing?" which I qualify as paranormal, show that they have a sort of shooting script of how they act in situations where a third tells them that artists do not have the copyrights. It proves that they know that the company isn't working properly. They ask: "How are you?" because they assume that it's not going well with somebody, a complete stranger to them, one of the millions of people who hear their music, because they think he/she suffers from immaterial damage, because copyrights aren't correct. They don't need to think that if they believe it is correct. That goes too far. Then they can better by behaving business-like and professional reassure the person. But probably they want to tape a conversation with that person, who will then as a result start talking about his/her vision in terms of copyrights, so they can use it later to prove to court that someone is crazy and/or harassing them. It is also not in their interest to wait until someone else goes to court with a copyright claim. They prefer to make the first step towards the court themselves, with preferably the confirmation from a security service that someone is troublesome. They don't want a case with solid copyright substance, and the legal system supports them in it. In short, the record company is, surprisingly, capable of grasping out into society to people in which they have interests. Moreover, certain people are at the same time sucked out and exterminated. This octopus shaped business structure they have, hand in hand with the image that they can make stars of ordinary people. That then seems to be more like nonsense, just as someone will unofficially not be crazy, if he/she disappears in an institution because of this.

Local observation

Once, I went physically to Sony Music for an observation. They are located in the media park in Hilversum. After having seen this, I can conclude that everyone who works there must be crazy. There are various reasons for that which every visitor should be able to imagine. The first reason is, because it is established in the media park. For a serious or major label it is of course not necessary to be located there, with its focus on broadcasting by television. That is a different industry. Record companies should be connected to it, but that doesn't mean litteraly be there. They can be dispersed throughout the country, but instead the major ones are all near Hilversum. In addition, the media park isn't normally accessible for everyone like elsewhere. Motorist can't freely enter it. Undoubtedly, there will be surveillance by security firms. So people who have no business there, according to its established companies, can easily be removed by the security services. This principle is quite puzzling, as these companies don't need to exclude people if they have no bad business intentions. Now the presumption arises that they are doing things there that are not allowed. On the other hand, one may wonder why it is all commercial and why it is not the government who monitors the media park. This for example to prevent criminals from taking over power. That is a possibility now, if only that criminal is commercial. Secondly, I base my conclusion on what I have seen in the record company. I walked around it and all I have seen are some large, wooden conference tables and a coffee machine on one side and on the other side an empty room with only a bright red desk. I have found nothing indicating that this is a company that knows about making good music. Probably they will have a studio on another floor with sound equipment, but the only thing of value must be their administration and that could be on a remote server. So for the external security combined with the fear of theft of their valuable assets, they do not need to settle in the media park. An average freely accessible rehearsel room company with twenty studio's filled with equipment will attract more attention from thieves. Someone whose music is being rejected by this record company, can't be ashamed of that with self respect. An employee of this company who visibly has to speak with so little knowledge, so much ignorance, who can't make music and with so little status, can't possibly accomplish this. Sadly, they probably have to hire an external musician, with good sense of music, to tell the people that their demo is not good enough. They are after all merely accountants. However, I believe that it certainly is possible for the staff in their business to at least do that with great conviction and a cup of coffee.

Videoclips

Making videoclips is another subject they do not want to talk about. Because of the doubts about the copyrights on music, it makes sense to also question the copyrights on videoclips. With video's you can film private business. Important is who devised it, how is was devised, what it is about, how clips are made and by whom. Probably that happens, as with producers, with external directors from outside the company who will then be enabled. But I can find no job openings for making video clips coming from record companies. Nevertheless, '*the making of*' programmes clearly show that for all those present it mainly seems to be a well-loved source of income. Video clips are often made elsewhere in a building or outside in the public area. Because public space is freely accessible for everybody, it is a possibility that people pass themselves off as a director or part of the film crew, actor or dancer. This is facilitated by a weak social control due to the fact that you only have to work once on one project, so you can remain sort of unknown to those involved. With a business suit, e.g. a T-shirt or a car with the name of a company or with imitated business cards, thirds may play a role there. Video clips are, just like cd's, flagships of the record company and they should be able to display a lot more knowledge about it.

Demos

The rejection of demos is another thing which is done a lot by record companies. One little mail to a another, smaller record company shows how difficult it is to get through to them. A musician who wants to speak to them about doing business, can hardly establish this. We have to put it on mail, because the receptionist knows too little about it. After no response is recieved, the receptionist says on the phone that it can be true, because they get 7000 emails a month. So on the phone there's nobody to talk to and the mail remains unanswered, because they get too much mail. As a result, the way they do business with new musicians is not sufficient. Yet, the damage is relative, because it is not necessary for musicians to work with a record company. Together with a band one can record music in a studio, put it on cd and self-release it. Many bands start out doing it that way outside of a record company. The costs of this are not to be underestimated. Hiring a professional recording studio runs into hundreds of euros per hour, and besides that the pressing of the cd's still needs to be paid out of the own pocket. Only after that the profit comes, should the cd enter the market.

7. Advocacy

Various law firms are contacted to obtain information about the legal aspect of copyright infringement in the music industry. They are googled through the internet and some of them were found in the telephone book. Some of the approached lawyers advertise that they are specialized in (mass) media affairs. Most of them are approached without explicitly saying that it was about a research. They were simply told that it concerned a copyright case in which the help of a lawyer was needed. From by far the most of them it shows on the phone already that they are not going to take the case. At that point no concrete information about who and what hasn't even been been mentioned, such as the band name, location, the text or music that has been stolen, the name of the offended party or the real author. When this is addressed anyway, they interrupt the conversation, they get angry, curt, change the subject of the conversation or continuously repeat during the rest of the conversation that they don't understand it. The latter is downright incomprehensible about someone who studied at least four years at the university and who is spoken to in the mother tongue. While they are approached about a topic that clearly has much overlap with their specialization; the law. If they do not get mad, they explain that a name registration of the song at a copyright organisation is needed to demonstrate that an artist has no copyrights. When two applicants come for a complete piece, **then the date of registration prevails** in court. There is no adequate argumentation that one of them has made it on the basis of written documents, recordings, adaptations or even witnesses. In the cases, lawyers indicate, that appear in court, so in the jurisprudence, both parties are performing artists. With only one, an in media specialized lawyer, I managed to make an appointment to bring the case to his office to discuss. It was probably successful by not going into the matter in advance on the phone and by clearly mentioning that **the case was sensitive.** In his office it was clearly explained that the music industry and artists had made an production, in which someone who has nothing with the music industry, was parroted, observed, with untraceable communication and interception equipment was tapped, that this data is processed, that the industry, using a database, puts itself in the place of people and their relations, while using mockery and sadism. Besides that, some of the already obtained correspondence with record companies and the survey data (see chapter 10) that contain the general opinion, was put to him. Unfortunately this was pushed aside. He also states, nevertheless, he is not able to do such a case and then comes with the same explanation as his collegues. But I could nota bene according to the lawyer go to the police if I think that people are secretly tapping information in my private sphere, because that is not allowed. In short, because judges apparently decide that a registration date of a song gives someone copyrights, lawyers will only take cases when someone can show an earlier registration date of the work. Thus the advocacy. Hereby essentially a scandal is demonstrated. Registering a number, of whoever it basically originates, obviously gives someone a lot of power. Even in the courtroom.

One explanation for the negative attitude of the lawyers in hindsight lies for the taking. In the past, they earned their living and (current) status by defending copyrights of artists with large and small record companies who claim their copyrights on the basis of the registration date at Buma Stemra. This of course cán not serve as evidence. Music is created like this as if by magic, songs come quite out of nothing. Some even come from nothing and are immediately a big hit and remain it for decades, without that seemingly anyone can change that status. In this way, lawyers became rich in the past and they still think to get rich from it. A repeat victim of the mass media continues to sit at home and knows about nothing. It is providing a default and lawyers can't be unaware of this. They know this and they are ashamed of this clearly for themselves, their colleagues and the administration of justice. Lawyers are in fact embaressed a lot when someone specificly goes on about a real copyright conflict. It is a case that would prove that earlier case done by lawyers, were not done correctly. There are now lawyers about whom information would reach the surface that they used to make money with dirty business, or the office they worked for, or they are negatively affected because of their naive believe in the law which is proven to be unfounded. In an attempt to maintain their status, they drill every real legal proceeding into the ground before it even begins.

Yet, every intelligent person whose interests are negatively affected by artists, record companies or whatever with illegal copyrights, initially has to begin something by legal means. There is another reason that can be mentioned why there are only limited opportunities for a private person in this case. Even with a registered sample, one can do little against for example a big (or small) American artist with a record company in the United States. Such a citizen has the general public opinion against him/her. This will reflect in the vision of the lawyer who will then not take the case, while copyright thieves can easily send their information via the ICT network abroad. Networking through the worldwide web is much easier than that a human being itself can travel. The level of non-understanding of professional bodies for this type of relatively simple actions is however extremely incredible. This despite the fact that they and especially the police, then play a major role in making this operate accordingly. They can, for example, easily tackle the publisher, the source of the problem.

In addition, the financial condition of the victim will often be insufficient to challenge a record company in a lawsuit. It also isn't one single record company, but a lot, because their policy shows a structural problem. Their method ensures that record companies do not know who did what. They blindly take everything that generates money, all the music has already been made long before they start working. It is possible to find a record company, but then it is a budget of for example €1.000,00 of the victim against €25.000,00 of the record company to pay the legal process. Financially seen it is ridiculously disproportional. Also it can be a case of one person against 100 people per song who all can show with a pay slip they are involved or want to prove integrity with a pay slip. So the continuity of the record company remains secured by the judiciary. Even when an individual can prove with specific evidence that he suffers from damage and that many record companies simply have to stop and should be written out of the Chamber of Commerce.

Because of this legal state of affairs a number of very reprehensible things remain unsolved. A victim that sees itself negatively affected by the music industry, consequently suffers from immaterial damage. This has a self-reinforcing process, because it concerns a mass medium that ultimately has its effect on many people who hear the music. On the one side a victim gets psychological problems; it drives you mentally nuts. It may be about private circumstances, which the victim probably had prefered to keep private. On the other hand, he at one point, also suffers from psycho-social problems. The victim is being harassed by the radio, which will always just be a background sound for other people, and from which people draw the wrong conclusions.

Record companies dealing with deejay music seem to have an additional reproach factor. Where in pop music various instruments and singing may still need mixing, deejay music usually is somewhat more monotonous, with few vocals and a computer-generated sound. Then the record company with all its means to make music is very bad, if they steal these '*beats*'. In that case they do not look after the self made creations of their deejays and producers. The music is easier to make. But how someone should do this, they apparently don't know. Thereby should also be thought of the uselessness of the record company, it is only a publisher, a booking office and a sharer in profits. The record company actually has no face, fans probably only know the company name and have an image of its image. The overall picture that people have in mind of a record company, then isn't true. The influence of a record company on making music can be nullified. What remains is that people can achieve something by pretending that they are from such a record company or that they are only once engaged in a mission of the company. They can, knowing who did or did not make the music, by corruption get acces to legal papers, such as labor contracts and pay slips, silencing both friends and enemies. They can generate revenue by working together with executing musicians and legal organizations like Buma Stemra, who don't control when, how and who are really the makers of music. While the people in the music industry make money, the real creators of the music are kept away and an aggrieved party is denied. It is damaging that lawyers don't want to start a legal procedure against this. It hurts a sore point. Now one can for example find people on a site who prepare a concert, distribute advertising or seize a space somewhere with professional material, printed T-shirts, etc., while people are kept away who don't want to benefit unfairly. On the one hand they work with legitimate organizations, on the other hand it gives them the knowledge, opportunity and experience to compile official papers, while it is not legally correct.

8. Police / Governance

The police doesn't maintain copyright laws. This is evident from a visit to two police stations, where I to be better safe than sorry, posed as someone with a band and as someone whose texts have been secretly recorded and used by others / the music industry. By this approach a reaction from the police was obtained, as in a real situation. When visiting the first police station, I explained that I was from a band and had recieved sheet music from unknown people. The content of the music was about personal, economic, social and political issues that were connected to the situation if I would act like it was my music. I even supposedly found a record company who was willing to sign a contract with me for that music. Because it is so strange and also sort of a private matter concerning the content of the songs and the fact that I will soon probably be approached a lot by strangers and because I don't want to do anything criminal, I ask the police wether they can do something about it. They immediately respond clearly that they can do nothing about it as if it is very far from their business. It is a civil matter and so the police will do nothing about it. I can freely go to the record company, sign copyrights and a Buma Stemra exploitation contract. When someone comes who does not agree with it, thát person can start a civil case. The police does nothing. They do make the suggestion that I might inform Buma Stemra or the record company about the problem.

At the second police station I explained that I came from a lawyer (which was true) in connection with a case on copyright. According to the lawyer, my texts were illegally recorded in my home and for that I could go to the police, according to him. There were three policemen at the desk of which one of them started talking past me each time. I was shown the door, because I should have registered it. I then clearly explained in several different ways that various offenses were a nuisance to me. That is simply the consequence of a copyright case. In addition, it can chase someone down, because it may be about confidential data. However, this doesn't affect the policemen at all. There is no understanding for it, because they even have to laugh and become brutal, because as far as they are concerned it is the end of the conversation. The failure of the police to act and the assessment of the situation as a mixed up civil matter is actually ridiculous. The quantity of people who all humiliate just one victim, makes it unmistakably a criminal case. Further it is imaginable that the victim is only aware of the problem if he/she hears the song. The song then is really just the tip of the iceberg, because the harm has already been done and exploiting it brings a psychosocial effect. There is nobody to protect the victim if anything goes, and if the victim then should begin a civil case against the offenders afterwards. Policemen can not possibly expect of the victim to always start a civil case, because it is an impossible task to know all music. Also, one apparently never wins a civil case, in which among other things is said that the artists are not the makers, without an earlier registration date of the same song, which of course nobody has. The musicians will think of something which makes it far too late for the victim to respond and principally will want to keep the music away from the victim to avoid that he/she recognizes something of him/herself. They will prefer to live and perform their music elsewhere. This theory is indeed confirmed by the fact that many artists live in Hollywood. This also shows related to its connectedness with the television world, that it is a persistent and structural problem. The effect of the television is that it can give many people in a short amount of time a false impression of the reality.

Yet investigations requested?

Maintaining copyrights is apparently no cup of tea for the police. Yet, government control on for example record companies is necessary, for various reasons. Firstly, an artist has his own company and makes on his own a lot of money. This is actually suspicious anyway, because many other people fail at it. Money in general also is a goal for which crime is the means. So in the music industry criminals can earn money through crime while maintaining status, because they work for a legitimate company. One is there figuratively under an umbrella of the government, because not a single investigation service searches something there. Income and employment depends among others on the decision how many cd's to make and on promotional activities of others. The record company and other players in the music industry don't benefit if the artist can't claim the copyrights, because they only want to make money. Based on that claim they take a piece from the pie. For many players it is otherwise impossible to make money, for example because of ignorance and mental weakness. Employees are presented music that they can't make themselves and start to assign it to other people (the executants), because it otherwise creates a psychological conflict. By also in doing business laying the copyrights with certain people and making it solid, they complete the circle for the outside world. For an individual with a conflicting, so to speak, *'half'* claim it then gets very hard to break this circle. A *'half'* claim does not count for them. Such a claim of a third may concern only the text (lyrics), the melody, the notes played by one certain instrument or decisive instructions. An important one is the claim of a third, because he/she finds him/herself personally affected by the way the music is made and how it is being exploited in the media. As artists themselves are often not accessible, it is logical that such a person then reaches out to the record company, but this research shows that they deflect this. So there is no solution to this problem.

Secondly, it is less likely that a minor shows up with a hit. Further inquisition should take place here.

Thirdly, the government should examine wether, because of the simplicity of unjustified claiming copyrights with official bodies and the lower face recognition of the other band members, besides the lead singer(s), there is corruption and identity fraud (a change of guards among the less known band members). This leads to stealing money from those who do want to work seriously in music. Then in everything that has to do with making, performing and selling music there is a sort of crime. Given the nature of the industry, the music industry with its flow of money can be wrong. It is common that doing business, making deals, bookings, promotion, connections with clubs and production activities are done by others than the artist. Having a manager who does everything can even go as far as that he even arranges the music for band members or individual musicians. In this way, someone who is not too smart, can be lured from the conservatoire into the music business. This person can then make music in a band, while there should be at least a reasonable doubt in him/her about the origin of the copyrights. When someone as a performing musician in this manner makes it his profession, without (being able to) check(ing) how the music making process exactly went, he/she must have psychological problems, with serious social consequences.

Fourth, no intervention shall ensure that there is no normal labour market and that the media is manipulated. Apart from that record companies shelter terrorism (terrorism is a broad term) they actually do nothing with talents who don't first deliver a complete product. They have no positions for text and music writers; on their websites no vacancies can be found for this. There is nobody principally there who draws talents in, potential artists (bands) would all knock on the door of record companies themselves, according an employee. The (demo) material comes in by itself and is judged. Self-reliance and working alone are therefore very important for success in music. Only with a complete product something is appreciated and can there be paid. A potential artist has to do everything alone or slavishly do exactly what someone says if they do an audition for something. The music industry more like a meat industry, especially inspects and disapproves of a lot of people. There is normal paying morale like paying for something that has been done. A selection of people with an inborn talent to sing, gets paid. Instead of this decision that is made by TV and media figures, proven economic activities should lead to an income. If they blindly exploit copyrighted work without any control, they run the risk of sowing hate and they are guilty of positive discrimination by laying down people with a genetically determined singing talent in large contracts. If a record company looks more like a publisher, because they have no know how of making music, then the government should treat them as such. It is common knowledge that should be known by people to avoid disappointment among other things. There is nothing wrong with a better business insight to go after potential ambitions. A better spread across the country would be more obvious for this, than the remarkable establishment near TV studio's. For record companies can do their work everywhere; their objective is not that of a TV studio. Also, the spread of potential talent will be divided across the country; it is probably dependent on population density. It is in itself surprising that many major record companies are located near Hilversum and that many artists who get face recognized live in Hollywood. Now there are in fact a few resources available to manipulate the media with. And should one conclude from the fact that most known artists are all with major record companies, that smaller record labels have distribution problems? For many artists started off at smaller record companies or are still there (with a smaller range). When it turns out that elsewhere located or smaller record companies indeed have distribution problems, the government can play a role to improve that. It also is the question why all famous artists are with a handful of big record companies. Based on what do they take those artists? Did the artists really sell that many records in the past? Are the statistics of the record sales actually right? It is just administrative data. Is there decided based on a number? Is it proved that the artists have so much talent? Major record companies often publish videoclips and radio stations often at the same time all play the same songs, based on the number of sold records. Cd's are quite expensive and a lot of people are living on a minimum income. Do they really buy that many cd's of the music that already is for free on the radio, in clubs and on internet for download? If people for example buy a cd once a year and do not recognize the songs on the radio, that are suddenly repeated five times a day, one or two things are probably not done fairly. Due to several reasons there could thus perhaps be a fraudulent bookkeeping and corrupted methods. If suddenly, multiple times a day, repeatedly the same new songs are played on the radio, that could be an indication that radio stations are involved in a scam. The music industry also consciously works with the same people in the media and makes a few artists have excessively much priority over other artists. There is no economic basis for that, looking at the quantity, diversity and quality of the whole range. It is more logical that the media have insufficient grip on artists who do not live near a studio. Cameras are portable, pictures and reports can quickly be send to the office by internet and phone and also journalists are mobile, so it will probably have more to do with the cooperation of the artists. In addition, the cause may lie in a bad functioning media, despite the opportunities and the research field that they have at their disposal. Also artists elsewhere, outside a TV studio or stage, can more quickly let go of things verbally that are not right copyrightwise, which the media refuses to publish. But if artists don't contribute in that way to generating their income and if movies can only be made in Hollywood, it may indicate that there is indeed a scam in the music industry. The same accounts automatically for the book keeping of Buma Stemra. For example, they calculate the profit based on playlists of radio stations. An instrument to measure what comes across the radio everywhere does not exist, so radio stations probably have to send these lists themselves. In addition, radio deejays often decide on their own what gets on the radio and one can wonder why, based on that free choice and personal preference, Buma Stemra calculates the income of musicians. In that sense, Buma Stemra is a dependant organization and it's the question wether the data entered in their information system is correct and fair. It is clearly a mistake prone system. A responsible, inspection organization can't completely deny this. Yet, apparently it happens structural with the police that with a certain bias, the existance of some logical cause and effect situations are denied. Even when victims personally report the situation at the police station. It is plausible that victims risk being arrested by the police and getting prosecuted, when they get rightly angry, without their report is treated at all. Moreover, the public prosecutor, lawyers and the judiciary can be guilty of abusing their power and forgery, because they, as a concerned party or accomplice, don't want the

85

whole thing to be exposed. Because of that someone gets duped even more for a longer period and the case disappears to be covered up.

Fifth, by the way the music industry works, the danger exists that human rights are violated of third parties who are not involved in music. The industry allows that people serve as a *'source of inspiration'*. A source of inspiration doesn't need to have any further participation in the making of a song. No by him/her produced sounds are outside of his/her knowledge recorded and processed in a song. They produced no phonogram. Other people have only, because of the music industry, talked about that person, an event or a crime that concerned this person. The inspiration source is ignorant of these facts. Conversations about that person, events and crimes can be set up and take place in order to make someone for a longer period of time a source of inspiration. The police may or may not have been involved in a possible crime. If this is so, they will probably not have been openly aware of a relationship with the music industry. The crime has then not been solved correctly and later the victim can experience even more material and immaterial damage from it. With the current ICT facilities one can at a distance discuss the situation created around someone at some point in time (also place spyware in a house or spy on someone through the phone or computer) and relate this to the present age, circumstances or his/her way of thinking. Other than music (lyrics) one can also make reality television of this. The social environment surrounding the source of inspiration will play an important role in that. This is sad. When the record company allows that all music and creativity is *'pushed'* in the business instead of that it is made there from A to Z, it leads to exploitation of people outside of the record company. It forms an attraction for those people in society who are eligible (because of certain factors, think of illegally integrated people, human trafficking, pedophiles) to do unpaid work for the record company. This means by using recording devices, musical instruments and text processing programs, to exploit people for inspiration. It is after all allowed and appreciated in society. How many labels and bands don't call themselves or their music *'psychosocial'*? The music industry and the consumer who buys cd's and visits concerts legitimizes everything. In the long run it also makes the record company and the artists dependent, because otherwise methods and processes are blocked, without this outside input. There are really no employed people in the business who possess the essential experience, knowledge and skills. Yet, there is a kind of secrecy regarding this, related to the connection with crime, copyright law and shame? The research shows that many representatives show forth a deviant morality and an antisocial mentality. This confirms the theory of the *'firm Octopus'* (see p. 64) and that they probably lash out to more or less a network that insults and humiliates people, violates copyright law, using the wrong means and is not afraid to use violence. Here certainly lies a piece of responsibility of the record company and other companies in the music industry. There is no transparency, openness and picking up professional things that are brought in doesn't happen, but however what can be found is a mean, sadistic, commercial character flaw. *'Inspirational sources'* constitute a risk

87

for the survival and the income of a record company and the entire music industry, including journalism, in connection with violations of copyright and privacy. There is also the danger that a '*source of inspiration*' becomes a necessary condition, when a record company apparently can't function on its own resources. The record company denies the existance of these people, even when they are confronted with them and if their very own business data clearly show that it is possible. The above shows that the music industry would benefit from getting rid of thirds, so they don't experience loss of face. This advantage would decrease if such a problem of the industry is well known.

(Ir)Responsibility and neglect of the record company

It is possible that artists come from the conservatoire and that they have the musical resources or a singing talent. They can can be brought up bad, drugged and have knowledge of copying and working on music. They can be discovered and followed by anti-socials and get stuff imputed, blackmailed, threatened, tattooed, and so on. This can be started at an early stage. Then there is a boundary between what is allowed and what not. Eventually it provides a lot of money to make music professionally, but one has to lie about the copyrights. This is easiest if the lie is maximized, uncompromising with every other, nice possibility. The crux of the story is that it is possible through the record company, music clubs and –media and a (corrupted) law system and authorities.

In principle, one must pay to legitimate beneficiaries. Without any ground record companies only pay out the performing artists and producers who personally take a song to their business. They do not discuss the case substantially with third parties who later indicate to be a holder, to pay out the part that third party is entitled to. However, you always have copyrights, even without registering it. So record companies award this unjustified to themselves. It makes it worse to not pay while you can. A third or a source of inspiration may as a lyricist or composer be entitled to a portion. The calculation of money goes wrong if other people classify themselves as authors. Criminals can also be holders of copyright. In fact it should be taken into account when calculating the profit. However, if it turns out that they are too antisocial and someone indeed systematically humiliate too much for it, insult, shorten human rights, run a sort of human trafficking business or something similar, then these holders of course don't need to be paid. But in that case the record company needs to acknowledge that those people made it. If they in nothing else than a sort of publisher's position can conclude that someone else makes the music that they exploit, they should look for these people, so they can refer to them better. The decision wether or not to pay can possibly be left to the judge. Probably they don't do so, because they want to distance themselves from anything that could infect their image. Also, many unregistered, actual makers could agree with it that they do not get paid, because they are aware that they are delinquent and offend other people. Besides that, so many people could be working on it outside of the record company, that by sharing the profit not much if it remains or it is unknown how to share it (see p. 104 about the making process). They derive their income probably from somewhere else, so they have sufficient financial capacity anyway. They could find that sufficient. People can have several different motivations and making music can be a means to an end. A society can be developed in which a lot of things are wrong. Confusion and opaque are sowed and the media gets filled, allowing corruption and criminal values to get the upper hand anywhere.

9. Other findings

The chamber of commerce

Two excerpts from the chamber of commerce are bought to see what is stated in it about record companies. Surprisingly, those mention in the company description a lot of verbs and some nouns that refer to a record company. Verbs like: produce, exploit, design, maintain, acquire, manage, provide, book, organize, mediate and engage, relate to nouns like: audio and video recordings, sources of information, pop concerts, music publisher, music and copyright, related rights, trademarks, patent designs, management, artists, musicians, tours, at home and abroad, merchandising-, sponsoring- and synchronization activities. This may clearly show the objective of the company, but the way this goal is achieved, is not clear from that company description. Strictly speaking, for example acquiring copyrights isn't even possible, other than making a work from beginning to end by yourself. Copyrights always remain of the author himself. So what it says, is not allowed! What is possible, is that the company signs an exploitation contract with an artist, assuming that this is the maker. They then bring cd's in circulation and calculate the sales and the like. If the artist is only an executant, it of course remains exploiting someone else's work. From the company description in the chamber of commerce it can't be concluded to whose copyrights it relates too. Also, for example producing sound recordings can just as well be a repetition of someone else's work. What it says, has all very little to say. Besides that, it's not clear who does it. Of one record company with no less than eleven trade names, it is said there are zero employees, while another firm is sole shareholder and director. Of the other record company is mentioned that there are just under 50 employees, and a few personal details of three directors. At least one of which is not autochthonous and therefore perhaps living abroad and not speaking the appropiate language. Sole shareholder is again a holding B.V.

An employee of the legal department of the chamber of commerce informs me on the phone that the chamber of commerce purely only registers companies. This registration says nothing about the methods of those who are enrolled. Hence the chamber of commerce has no legal possibilities to intervene if the methods of an enterprise are in fact not correct. The mayor grants an exploitation permit and further the employee believes that one can go to the police for a civil procedure against an organization. This probably with the idea that the police should proceed to check wether there are criminal offenses committed within the company. As to that, the chamber of commerce wears blinkers. They refer back to the moment of enrollment. If back then there was no reason for doubts around the accuracy of the provided information – as if the data in their system is always right(!) - , no further investigation takes place. These are not actual data, but historical information of perhaps twenty years ago. A field service is present, they can e.g. after a concerned party wrote a complaint letter about a company, do a small research at the company. However, this is just a superficial business check, in the sense that for example a kebab joint c.q. record company should look like a kebab joint c.q. record company and it's good. As long as this is the case, a lot of crazy things can happen in that company without that the chamber of commerce ever removes the registration of the company from its system.

CD Presses

Record companies often outsource the pressing of cd's to cd presses. This was a reason to ask a few presses wether it was possible to get a tour. Unfortunately this was not successful. For many presses it lies precarious, because they also make software cd's. Potential customers or groups usually do get a tour. It is strange that they take the word precarious in the mouth, but not because of possibly faulty copyrights. It's clear that one can also easily make music cd's at home, with a burner on the computer. Cd's are inexpensive and are available in many supermarkets, just like microphones and other recording and playback devices, necessary to put selfmade music on cd. A cd booklet can also easily be made by a printing-office. Only when one wants it to look professional, the cd itself needs printing. One can go to some cd presses for this, who for example also print an image on blank cd's, in a limited edition, at a reasonable price. Conversations with staff from presses show however that many presses don't offer that option. They use a fairly aggressive selling method, while the income, especially of a starting band, is far from certain. Many companies only want to do the whole process (put the music on it, print the cd, make the book and the box) and work with minimum numbers of cd's, typically 500 pieces. Strange, as many starting bands should have problems with the costs of that. Especially since they can do a lot of it at home and at far lower cost.

Live music clubs

Via the internet a number of Dutch and American live music clubs were found and it was repeatedly tried to contact them by phone. However, it turned out that only one in six is accessible. These are clubs that have fixed arrangements with record companies. Some are *'member only'*; visitors must be a member of the club. For a fee, they can go to a live performance of a few bands (that are part of a label) of a record company, usually a couple of times a month. A lot of advertising is done for the music and that it is for music lovers, but there's essentially a strict separation biased between the visitors (fans) who just have to listen to the music and the bands who like (excelling?) booked specialists make the music. It is in itself a little strange, because that distinction between between listening and making music might be less sharp for someone who loves music. Yet this train of thought of the club manager and other people apparently continues, shown by the products that are sold at concerts. There is band promotion material, merchandiser T-shirts, posters and such available, but no guitar parts and other items to make music. Only one American club appeared to be structurally well accessible and they were therefore over the time span of several weeks more often called. However, it didn't add much, because the two employees were not really open for research and because one of them despite all efforts, tried to answer all questions by asking how he could help me. The other employee suddenly said after about the third phone call that the police were looking for me. The conversation shows that they believe that TV and sound can not be shocking, that they don't broadcast anything internationally (although the shows in their clubs can be seen world wide on Youtube) and that bands can't be shocking. The employee came about a bit stupid and untruthful when he went on to acknowledge that even someone in my own house could not scare me. In any case, then don't know how in their business, they can shock someone. On further inquiry, they make it known that they haven't been to school and have had no specific training for the music industry. One can imagine that a training is useful, e.g. because of the insanity of the impunity and lawlessness around internationally expressing insults. The music industry namely has, because of the way it functions, a window open to a wide international audience. In doing so, they occupy a significant proportion of time and space of various mass media. Band members are always allowed to just say something, while it is quite possible that it concerns someone else's private life.

Someone from a Dutch live music club was found willing to participate in a morality test. Hypothetically seen, these people should score low on this, if for example there are songs, that secretly are about someone else's private life, which is unethical. Homosexuality was chosen as the topic for the test, because it is not uncommon for artists and because it quickly shows wether someones morality is right. It is common knowledge that a homosexual shouldn't conduct heterosexual behavior. Also, not being able to maintain one relationship with a partner is an important psychopatic (antisocial) feature. It is plausible that there is psychopathology in people who wronlgy appropiate copyrights to themselves or someone else. The presence of such people with psychopathic traits and e.g. drugs may cause a forced lying in other non-psychopathic band members / artists.

Question 1:
"Do you think it is OK if your club would do business with an artist who in the media says he/she is homosexual and sexually abused as a child, but in real life, as it appears to you, has heterosexual sex with various people?"

Answer to question 1: Yes.

Question 2:
"What do you think is the most artistic thing to say for an artistic lesbian?

    A. I want to draw a boob.
    B. In two weeks a new song of mine will be released by the record company."

Answer to question 2: B.

The respondent indeed scores low on the morality test. To question one he clearly had to say "No" and to question two "A", because answer B can't proof that the lesbian was so artistic to create the song herself.

Tattoo shops

Many artists have tattoos. Tattoos label someone faster as an artist. These are two reasons to approach tattoo shops for this research. In Rotterdam a random tattoo shop was visited. In there a tattoo artist was observed from whose physical behavior it appeared that he was highly motivated to put a big tattoo, but probably because of the influence of specific drugs. Several tattoo shops in Hollywood (Los Angeles) were googled. An employee of a shop was willing to participate in a small telephone interview. It was a shop that in the last two years always had two or three employees (tattoo artists). Wether this is average for the industry is unknown. The administration and reception they do themselves. In the past few years the location of the shop has not changed. Tattooing only takes place inside the shop. Indeed, they also tattoo famous people. They have fixed prices so these famous people don't need to pay more. The price depends on the size of the tattoo and the like. Theoretically it could be argued that artists should pay more, because a tattoo contributes to their fame and thus income, but apparently this is not the case. The man denies that they have settled in Los Angeles due to the fact that many stars live there. It is unclear wether they would earn as much money in another area. Anyone aged eighteen years and older may come for a tattoo. According to the respondent artists who come, then talk about where they come from and about everything they are doing, like for example their (charity) projects and political role. So basically this says nothing about how they achieve their ambitions and the form, situation and steps in which they make their music. As the man in fact firmly denies by this that artists explain in his business how they get their copyrights, it makes it less likely that they are holders. After all it can be assumed that someone talks about what he/she is doing and that a copyright holding artist in his/her own time is busy a lot with making music. Asked wether he had answered all questions correctly or not, the man however openly admitted that he probably lied to protect his business and because of unfamiliarity with the interviewer. The respondent indeed openly said nothing negative about his business or about artists, although the research of course did have its focus there. Just like other people in this study, he occasionally gets suspicious, he refuses to talk or lies and plays nice weather.

On internetforums on tattooing it appears to be the norm to for no particular reason exhibit unrestricted hostility, anti-sociality, distrust and a lack of respect towards socalled '*homeprickers*'. These are people that tattoo at home without a permit. A permit is in the Netherlands available through the GGD, a social health service. After a fee is paid, one of their inspectors checks wether the applicant has a studio that does justice to the appropiate hygienic norms. However, on tattoo conventions several dozens of people get publicly tattoed in little stands. This clearly raises questions concerning the hygiene. By letting them use hand alcohol, the GGD allows it. On the internet all sorts of tattoo machines and other needs are easy and cheap to get. For that reason it can be assumed that there are many homeprickers anyway. It looks like there is hypocrisy in the game. Because of in fact initiating the secrecy and confusion towards 'illegal *homeprickers*'. This is because anonomous homeprickers probably work a lot in secret, besides making some black money. Perhaps they place, while travelling about, a lot of the same tattoos, which is for an artist identity very attractive. With the same physical characteristics as someone else, one can pretend to be that person and have themselves edited in a videoclip by someone with the right ICT facilities. This can happen quickly, but also long before anything is published. The memory card of a camera can after all be stored for years, before someone cuts it in a videoclip.

Internet

On the internet some, especially American, artists were followed. This means that various websites, interviews and performances were watched. The story of the artists on their music is often very vague. They give a general, unspecified meaning of a song. They mostly just want to be liked and promote their tour or cd. On the one hand, bands display a charity that makes people soft and on the other hand they often say they use their music to spit their bile about something, but that is only because they are artists and they need their freedom for it. Why, for example, should a man get angry if such a person says that he is against Bush, the president of America, and and blackens this man in his music (lyrics), because this singer or band is against warfare? Isn't that no more than an idealism that can make people happy? Or if an artist with his hard-earned money connects his name to a charity? Every band or who or what one is, only wants to make the world better. That is eventually, in the final analysis, the reason why everyone should like them. Don't we do that, then simultaneously in one breath we are treated negatively. Then initially we are inconspicuously threatened and we must leave, also according to the hired security, and even the government then plays a role. That government lurks on us and will stigmatize us, because we are (especially) psychologically not alright. Strange, that this costs a band so much energy, too. You don't expect these activities from labels and bands that you adore. A pity only that the issues raised by the artists related to their songs, can't be recognized in the lyrics or the music. Perhaps it can, if someone actually puts the word '*Bush*' in his lyrics once. Outside of that, it remains a far too vague and general story, anyway. Plus, no one can deny that sometimes, while listening, everybody sees a detailed description in a song of a moment of one or two seconds, an act, someone or of himself. What therefore has nothing to do with the problems of the world, that are going on for years. What if you are that person, whose life is immortalized in music? It's just a taboo. While it can be very nice when someone does that about someone in his environmen that he knows and also acknowledges it. But if there is a breach of copyright, this familiarity of the artists with the subject of their music will not be there. This has to do with the reference frame of those come up with the lyrics and musical notes. This takes place at the beginning of the making process of a song. But meanwhile, people just continue in and out of season to listen to music. The quality of music is also basically suitable for it.

On the internet it seems that on several discussion forums and chat rooms the IP address of the computer of people with insolent language or critical comments about the music or the music industry are permanently blocked. This happens often without a warning and a statement of reasons. It also happens that they are bullied in lengthy discussions by the other users when they are still present. This increases the likelihood of brutal language and likewise of being electronically banned by a moderator. A moderator essentially is is an unknown person who after logging into the website has the power to remove other users. The use of dirty words in itself is allowed. Thus, a lot of enlarged, personal stories are met about sex with people who are much older or younger, sex in general, sudden gay sex, sex with somebody else's girlfriend / boyfriend, suppressed homicidal tendencies, cheating as a normal phenomenon, kooky pick up attempts, etc. This can especially be found with great regularity in the metallic, rock music segment. This, while people who criticize it are quickly removed. This elimination of certain IP addresses and in return the rendering of in fact deviant attitudes towards life, indicates that the derailed morality, that some people in the music industry clearly show of in this study, continues on to relevant websites on the internet.

Psychological

It actually goes beyond the scope of this study to examine the psychological effects resulting from copyright infringement as previously described. However, one can say that there is a negative psychological effect when music continuously, in the same way stimulates the memory of a person (inspiration source) ánd no one acknowledges this. In addition, the spreading of such sensitive information through music will have psychological consequences for other people as well.

Festivals

A few visits to festivals provided the information that the real music lover has a big chance to miss much of the offered live music there. At some festivals, namely, there is simultaneously performed on ten stages at the same time. An estimation based on a visit and the program schedule of the Black Cross Festival in Lichfield shows that one can only attend to a maximum of 25% of the offered music. Apparently, it is merely aimed at thousands of people to come for a beer and a snack with the music in the background. Nobody thought of it that a music lover could care less and maybe prefers a smaller-scale performance of his favorite band. Besides rock T-shirts, no goods for use in music are sold. Also, there are no relevant companies present to expand the network of a potential rocker. It is unclear wether this has to do with a discouragement policy towards visitors and a degradation of visitors to fans, whereby a T-shirt then again is an upgrading of a fan. In any case, it is in a bright contrast with for example an equestrian event, where always a lot of relevant goods are sold and companies present themselves. Because of the explorative aspect of this research, I take the liberty to indicate that there also was a pro-pedophilia concept in the form of a piece of land with children's attractions, which as a whole was in fact only accessible for children. Adults were only allowed in together with a child, regardless of wether both know each other. Many security people controlled on this among other things, without that the responsibility for one or more children was crucial for letting an adult enter. It would be more logical to open the area normally and just make the attractions exclusively for children.

Guitar lesson / Rehearsal rooms

Some rehearsal rooms are visited and used for solo playing on an instrument. In addition, a guitar lesson was taken in the course of the research. A rehearsal room usually consists of a windowless room with insulated walls and doors, a drum set, a number of (guitar) amplifiers and a mix panel. Despite the isolation it can be well heared what is played in another room, when one rents a room. Based on observation, there appeared to be a lot of bands compared to soloists. Moreover, the focus seems to lie on repetition and not on making or improvising something new, which is also possible for a band that writes its own music. Strikingly, the music from the other rooms sounded much better

103

than the music produced by myself. The difference was even evident e.g. for the drum set by the fact that the sound of a strike sounded above average with a band from another room, while it sounded less that average with me. Presumably, there is therefore being manipulated with the sound in rehearsal rooms. Because this is probably going through the amplifiers, this would mean that they are manufactured in such a way that this manipulability can be build in. It may be possible for example that the machine plays interfering or improving sound files at the same time. It is then logical that this probably also can be controlled with a kind of remote control. In any way, for a third there is much opportunity to make sound technical adjustments in a rehearsal room. Because the room can be rented, without supervision over what happens in it. It makes however more sense if the cause already lies somewhere in the production process of the equipment. Scratches and such would otherwise prove certain actions done to it. If this is the case, and these adaptations are there, then it could of course on the one hand lead to it that someone prematurely ends his music career. On the other hand, it could accelerate the process for others, because there are things that happen by itself. For a layman the electronics in music equipment is too complex and inaccessible to check its proper construction and functioning. At every performance or festival with amplifiers it is therefore theoretical possible that someone in the audience

with a remote control operates the sound. This is in fact no different from wireless surfing the web, or operating a home cinema set or another system at home with a remote control. It is however to be recognized that the activities of those on the stage somewhat have to correspond with the sound. The person or people operating the sound or giving instructions can therefore best be found in the audience.

From meeting the bands that were present in the rehearsal rooms, it can not be said with certainty that everybody of them is voluntary there. It is well possible that an offense is committed   against certain people, because systematically, some individuals seem to be very down. Some people, on the opposite, seem the walk away with certain things, while others hold an agressive attitude that can't be explained.

During a lesson, the guitar teacher offered some more insights into making music. So, improvising a song is easier if the band members already know each other for a long time and it was confirmed that a songtext can indeed be written separate from the rest of the song. All the instruments of a song, lyrics included, can together be noted in sheet music. The most common instruments in pop music are: drum or percussion (drums), bass guitar (usually two), guitar (also usually two) and the piano. That means for a band that there is usually a drummer, a guitarist for the melody, a rhythm guitarist and a singer who often has a guitar as well.

This, and among other things, listening to music and some practice led to a theory about a working method that makes it possible for a few collaborative, specialized people to produce a large number of songs about a limited number of topics. The theory is based on the assumption that instruments, beats and text are essential for a song. The theory, however, makes it less likely that one person in one day writes a song alone. This has to do with the actions that are necessary for making music. In this theory, these are approximately in chronological order the following actions:

1. The recording of texts.
2. The recording of beats; tunes or melodies.
3. With help of a computer with a word processor store and arrange the beats and texts.
4. Try to play the beats on various instruments and write down the notes.
5. Categorize the texts, translate them and make lyrics with stanzas.
6. Picking and uniting the obtained material to get complete songs.

This allows someone with a large file with notes and lyrics quickly and easily to come up with a new song or cd, if necessary withing one day. One can also in principle do this apart from the performing executants, while it may concern someone's private life, who is kept out of everything. It is conceivable that people work together in this via the internet and that each one will probably specialize in a specific function on the aforementioned list or on a musical instrument, for a fast and efficient workflow. The yield of the theory gets even clearer using the following arithmetic sum with a indicative formula, for 3 instruments per song:

Number of songs = (Number of beats) (Number of possible musical instruments)^3 (Number of texts)

Take for example 3 beats, 6 available musical instruments (3 per song), and 1 text. That is 3*6*6*6*1=648 possible songs. This fomula serves of course only as a tool to give somewhat of a numerical indication on the fact that based on the same data as a foundation for music, one can make several songs. One can also imagine that a beat, or a tune can be played in several ways on one instrument. For example a drum set can have a dozen items to strike, while a guitar has many more elements. So actually the possibilities seem quite endless, despite that the number of possible instruments and the number of instruments per song have a more or less fixed value. If the music industry lends itself for this, then certain people are at risk of being excessively blackened. Because people can just be doing it, it is also logical that when the data is about a person, that person will experience some discomfort in dealing with the social environment. For that, the music industry is also responsible.

10. Survey data

A small-scale survey, conducted in 2009, on the public opinion about popular music (see appendix), shows a general tendency among the population to find it a real possibility that artists are unfamiliar with the makers of their song. If artists unlawfully openly claim copyrights, then according to a lot of people this doesn't matter for the sale of cd's and concert tickets. Besides that, many people do not know, based on the facts and information known to them, wether the artists they know, write their own music. The survey was conducted among 64 respondents aged 13 to 85 years. Half of the respondents is male and the other half is female. The average age is 38 years, taking into account that the age of three respondents is not known. Two respondents were still a minor at the time of conducting the survey. 22 People were contacted by phone using a phone book (of Rotterdam and environs). The rest was asked to fill in the survey on the beach at Scheveningen. The questionnaire, consisting of 21 questions including sub-questions, did not lead to useful results for all questions. Some questions referring to gossip papers were not answered by sufficient respondents and led to unreliable data. Therefore, the research leaves those questions further out of consideration.

Table 1 shows six of the survey questions. Regarding question one, the results show that two thirds (64%) of the respondents would mind it if artists say they write their own songs, when in reality that is not the case. One third (36%) has no problem with it if they would do this. Question two then shows that again two thirds (64%) believe that it doesn't matter for the sale of cd's and concert tickets if the artist in public claims his copyrights, while he/she is not the maker. More than one quarter is negative and suspect that the sales then would get less. Two thirds answered 'yes' to question three. They consider it a real situation that an artist doesn't know who made the song that he/she is publishing. One third does not find that a possible situation. About question four the opinions are divided. About half of the respondents don't know, based on facts and information known to them, wether the artists that they know, write their own music. The other half thinks they do know it, based on that. As for question five, table 1 shows that over two thirds does not think it has to do with censorship, if it says in publications on popular music that artists write their own songs, while that is in reality not so. One respondent added it has to do with money and another called it *simply lying*. Question six shows that more than half of the respondents does believe there is some censorship in publications on popular music. More than one quarter believes this is not the case and a small part of five people doesn't know this. Censorship is defined in this study as *any kind of supervision on printed matter intended for publication*. This definition leaves some margin for e.g. control in a civil or commercial manner by the social environment and by individuals besides government supervision.

Table 1

Questions	Less	Doesn't matter	More	No	Yes	I don't know
Results of a small scale survey about opinions on pop music among 64 (N) respondents aged 13 to 85, sex ratio 1:1						
1.	-	-	-	36% (23)	64% (41)	-
2.	27% (17)	64% (41)	6% (4)	-	-	3% (2)
3.	-	-	-	33% (21)	67% (43)	-
4.	-	-	-	48% (31)	52% (33)	-
5.	-	-	-	70% (45)	30% (19)	-
6.	-	-	-	37% (24)	55% (35)	8% (5)

1. Would you mind it if artists say they write their own songs, while they don't?

2. (See question 1) Would an artist because of that sell more or less cd's and concert tickets or doesn't it matter?

3. Do you think the situation is possible that an artist doesn't know who made the song that he/she is publishing?

4. Do you know based on the facts and information known to you, wether the artists that you know, write their own music?

5. When publications on pop music say that artists/bands write their own music, while that is in fact not the case, does that have to do with censorship according to you?

6. Do you think there is censorship at all in publications on pop music?

Table 2 shows how the respondents think an artist is stopped from publishing a song, if he has it (e.g. sheet music and samples) in his possession, while the artist does not know who made it.

Table 2

How is an artist stopped from publishing a song, if he does have a song in his possession, of which he does not know who made it?	
	Percentage (of N=18)
I don't know	28% (5)
BREIN/Buma Stemra	33% (6)
Lawsuit other artists	11% (2)
Not	11% (2)
Not interesting	6% (1)
Manager/Team	6% (1)
Connoisseurs	6% (1)
No means	6% (1)
Record company	6% (1)
Total number of answers	20

Because of the design of the questionnaire, this table only contains eighteen respondents who indicated at question three (table 1) that they consider it impossible that an artist doesn't know who made the song that he/she is publishing. It then turns out that over a quarter of those people do not know how an artist is stopped. One third believe that these artists are stopped by so-called copyright organizations such as BREIN and Buma Stemra. 11% Of them think that they are put to a stop through a trail of the other artists who made it. Another 11% think they are not stopped. Furthermore, the respondents mention the following possibilities: by the manager or the team, the connoisseurs, the record company and because the artist has no resources (money, producer or label). One person thinks it is not interesting for an artist, because the profit does not outweigh the cost. Strikingly, the respondents here seem to disregard the fact that the parties named by them, may have financial and other motives to let the artist then publish a song anyway. Despite that, they still come up with some interesting answers.

Table 3 contains a sub-question that was asked only to respondents who reported, based on facts and information known to them, to know wether the artists they know of, write their own music (see question 4 in table 1). When asked on what facts and information their knowledge is based, they were free to give multiple answers. This resulted in 41 responses that are grouped in the twelve categories in table 3.

Table 3

Based on what facts and information do you know wether the artists that you know, write their own music?

	Percentage (of N=30)
CD (case)	40% (12)
Showing interest in the artist	3% (1)
Interview	10% (3)
Media (news)	10% (3)
Active in the music industry or familiar with such a person	7% (2)
Hearsay	7% (2)
Read it (newspaper/magazines)	27% (8)
Belief/faith	7% (2)
Meeting the artist	7% (2)
Common knowledge	7% (2)
Internet	10% (3)
Intonation of the artist	3% (1)
Total number of answers	41

The majority, 40%, refers to the cd. On it, or in the case, are according to them the details of those who made it. More than a quarter says to have read it in newspapers or in magazines. The other categories that are mentioned to a lesser extent are: showing interest in the artist, interview, media (news), active in the music industry or familiar with such a person, hearsay, believe/faith, meeting the artist, common knowledge, the internet and the intonation of the artist.

Finally, table 4 shows how probable people find it that one person or a band, without outside help, makes his/her/its own songs.

Table 4

Probability on a 5 point scale of making songs / cd's by one person or band alone, without outside help. (N=64)	Applied: 1=highly unlikely 2=a bit unlikely 3=not likely, not unlikely 4=a bit likely 5=highly likely.
Questions	*Average*
1.	3.27
2.	2.16
3.	2.94

1. To what extent do you consider it likely that one person or a band (as a unit) writes a hit (song), without outside help?
2. To what extent do you consider it likely that a 15 year old child writes and publishes a hit (song), without outside help?
3. To what extent do you consider it likely that one person or a band (as a unit) writes and publishes several cd's during a career, without outside help?

The degree of probability was measured with the numbers 1 to 5. It is considered the least probable that a 15 year old writes a hitsong with an average rating of 2.16 and the qualification '*a bit unlikely*'. It is found somewhat more likely that someone or a band writes several cd's in a career. This has an average of 2.94. That a band or an individual on its/her/his own writes a hit song, scores the highest on the scale of probability with an average of only 3.27. On the whole, the results show quite a low average. From that can be concluded that people actually don't (also not highly or a bit) find it likely that artists or bands make their own songs. The public opinion is therefore contrary to how copyrights are often officially registered.

11. Conclusion

To sum it up, now a few things can be said following the research questions that were formulated in chapter two. These were the following questions:

1. *"Are the registered copyrights in the music industry right?"*
2. *"How do companies and institutions in the music industry function?"*
3. *"How are songs made?"*

The research shows that sub-question three is answered in different ways. While a music teacher can explain this, record companies seem to have no knowledge on this at all and artists behave mysteriously concerning it, they develop according to the court as if by magic when Buma Stemra registers them. Therefore, a plausible theory is formulated with a making process of a few steps, that can also be completed outside of the record company by a social network.

Question two is described in detail in the various chapters of this book. A general trend is apparent that many individuals in the music industry display an antisociality and a deviant morality. So it is apparently considered normal by many players if the media for example of a heterosexual artist would say that he/she is gay. The image of record companies that they have much knowledge of making music, appears to be incorrect. They are especially a body that judges music demos and selectively promotes music. Record companies, Buma Stemra, music publishers and the representatives of copyright holding composers work like book-keepers, without checking wether copyrights are correct, not even when someone requests it. This is explicitly stated by them. One record company even became threatening when suspicions arose that they were confronted with a copyright claim of a third. In case a third lays claim to a copyright, a record company does not negotiate. In that way, they keep troubles away and they announce that they don't want to be harassed by someone who is no partner. A third will have to start something by legal means, but for a private person there is little opportunity for that. As it appears, many lawyers don't even take such a case on.

From the above, it already appears that the first research question can be answered with that a copyright infringement in music can actually happen, because it is structurally possible. Also, systematically important information about copyright is missing, such as how something was made and what exactly it is about. This can be deduced from: magazines, journalism, the language of business partners, the lacking of contact information in crucial publications, the story of editors and the story of artists themselves. Artists are usually very vague and speak in general terms about the meaning of a song. They mostly just want to be liked, talk about their success and promote their tour and cd. Critics are excluded from relevant websites. Editors have, as regards copyrights, no substantiated story either. Often, no sources are mentioned in magazines and texts and there is no chamber of commerce number listed in record companies' websites. In addition, the attitude of companies in the music industry is on the defensive and not understanding anything, although no

Chinese language is conducted. Since they are not susceptible for certain (emotional) issues, somewhat psychopathic people appear to be working in the music industry. They don't spread any concrete information regarding their practices, draw up a wall, do not understand it, declare people crazy and cover cases up. This, while infringing copyrights can lead to psychological damage in people. It is e.g. easily possible for artists to, for money, quote things from other people in the media, without saying of whom, as if it is a statement of themselves. This is in fact a kind of plagiarism. The involved record company and the media will then have to intervene themselves in order to maintain copyright laws, instead of just wanting to earn money. The research shows that this mechanism is not there. On the contrary; the company description in the chamber of commerce seems to go against the grain, concerning the law.

A few things in this book may look like excessive criticism here and there on the music industry, but yet it can be founded on objective scientific research, which therefore can be confirmed by a repeat examination. The statements in this book are well substantiated. Because of the methods of record companies and Buma Stemra there is a chance that something isn't right. Also, certain circumstances point out that indeed something is structurally wrong, like the essentially useless establishment of record companies and artists near TV studios. For video and audio material can also via mail or the internet be sent to a TV studio. Further, a support of the advocacy and the police is demonstrated for the music industry and a refusal to deal with certain copyright crimes. This allows anyone

118

who wants to, to commit a crime concerning copyrights with impunity, without that any negative consequences follow. With regard to third parties, record companies apparently have the judiciary in their pocket. The fact is, judges decide based on nothing more than the three words of an artist: "I made it.", and the registration at Buma Stemra, that the artist has made it and not based on someone else saying the same thing based on his own administration..

Thanks

I would like to thank everybody personally for their coöperation that made this book possible. Because of the critical content and the protection of personal details of those concerned, this will do; without the mentioning of individual names. Also, that would be unfair to the probably much larger amount of people whose name I don't know, who made an effort.

Resources

BREIN foundation

Buma Stemra (comparable to the RIAA in the USA)

BV Pop

Chamber of commerce in: Rotterdam, Amsterdam and Gooi-, Eem- and Flevoland

FIOD (financial information and investigation service)

GGD (dutch social health service)

Intellectual Property Hotline

Police: two offices in Rotterdam

Record companies:

      AT Productions

Bug Music
Cardiac
EMI Music Publishing Holland BV
Koch
Midtown
Peer Music Holland BV
Red Bullet
Sony Music Netherlands
Universal
Warner Brothers Music Holland

Tax Rotterdam

Websites: Allmusic.com
          Forum.tonsoftattoos.com
          Myspace.com/music

Search engines: Youtube.com and Google.com, for searching bands, interviews, shows, websites, lawyers, magazines, record companies, live music clubs, tattooshops, rehearsel rooms, etc.

All sources are dutch, except some of the record companies and the sources from the internet. Those sources are mostly connected to the United States of America.

Appendix: Questionnaire public survey

1. Do you have any journalists in your family?..yes/no
2. Do you have any musicians in your family?..yes/no
3. Can you mention a few pop artists? (mention 3)
   ……………………………………………………
   ……………………………………………………
4. Can you mention a few dutch pop artists or Idols? (mention 3) ……………………………..
   ……………………………………………………
5. Do you read magazines about pop artists?
   ....yes/no

Skip question 6 if you answered question 5 with a 'No'.

6. a. Do you think there regularly is gossip in magazines about pop artists?....................yes/no
   b. Do you think the amount of gossip is unacceptable much?................................yes/no
   c. If there was gossiped like that about you in the magazines, would you find that unacceptable?……………………..……..yes/no
7. Would you mind it if artists say they write their own songs while they don't? ....yes/no
8. (See question 7) Would an artist because of that sell more or less cd's and concert tickets or doesn't it matter?..........................More/Less/Doesn't matter
9. This question you must answer on **a scale of 1 to 5 of probability**. The meaning of **1 is highly**

unlikely, 2 is a bit unlikely, 3 is not likely, not unlikely, 4 is a bit likely and **5 is highly likely.**

a. To what extent do you consider it likely that one person or a band (as a unit) writes a hit (song), without outside help?

1   2   3   4   5

b. To what extent do you consider it likely that a 15 year old child writes and publishes a hit (song), without outside help?

1   2   3   4   5

c. To what extent do you consider it likely that one person or a band (as a unit) writes and publishes several cd's during a career, without outside help?

1   2   3   4   5

10. a. Do you think the situation is possible that an artist doesn't know who made the song that he/she is publishing?.................................yes/no

Skip question 10.b. if you answered question 10.a. with a 'Yes'.

b. According to you, how is an artist then stopped from publishing a song, if he does have a song in his possession, of which he does not know who made it? ....

....

....

11. a. Do you know based on the facts and information known to you, wether the artists that you know write their own music? .....yes/no

Skip question 11.b. if you answered question 11.a with a 'No'.

b. (See question 11.a) Based on what facts and information do you know that? ....

123

....
....

12. When publications on pop music say that
    artists/bands write their own music, while that is
    in fact not the case, does that have to do with
    censorship according to you? ....yes/no
13. Do you think there is censorship at all in
    publications on pop music?.......................yes/no
14. What is your age and sex?...........................M/F
15. In what field are you employed?........................
    ....

Do you have any questions or comments about this
research?..
....
....
....